CONTEMPORARY MATHEMATICS

Titles in this Series

VOLUME 1 **Markov random fields and their applications**
Ross Kindermann and J. Laurie Snell

VOLUME 2 **Proceedings of the conference on integration, topology, and geometry in linear spaces**
William H. Graves, Editor

VOLUME 3 **The closed graph and P-closed graph properties in general topology**
T. R. Hamlett and L. L. Herrington

VOLUME 4 **Problems of elastic stability and vibrations**
Vadim Komkov, Editor

VOLUME 5 **Rational constructions of modules for simple Lie algebras**
George B. Seligman

VOLUME 6 **Umbral calculus and Hopf algebras**
Robert Morris, Editor

VOLUME 7 **Complex contour integral representation of cardinal spline functions**
Walter Schempp

VOLUME 8 **Ordered fields and real algebraic geometry**
D. W. Dubois and T. Recio, Editors

VOLUME 9 **Papers in algebra, analysis and statistics**
R. Lidl, Editor

VOLUME 10 **Operator algebras and K-theory**
Ronald G. Douglas and Claude Schochet, Editors

VOLUME 11 **Plane ellipticity and related problems**
Robert P. Gilbert, Editor

VOLUME 12 **Symposium on algebraic topology in honor of José Adem**
Samuel Gitler, Editor

Titles in this series

VOLUME 13 **Algebraists' homage: Papers in ring theory and related topics** Edited by S. A. Amitsur, D. J. Saltman and G. B. Seligman

VOLUME 14 **Lectures on Nielsen fixed point theory**
Boju Jiang

VOLUME 15 **Advanced analytic number theory. Part I: Ramification theoretic methods**
Carlos J. Moreno

CONTEMPORARY MATHEMATICS

Volume 14

LECTURES ON

Nielsen Fixed Point Theory

Boju Jiang

AMERICAN MATHEMATICAL SOCIETY
Providence · Rhode Island

Some material occurring within will appear in 1983 in the Pacific Journal of Mathematics under the title "On the Computation of the Nielsen Number".

1980 *Mathematics Subject Classification.* Primary 55M20.

Library of Congress Cataloging in Publication Data

Jiang, Boju.

Lectures on Nielsen fixed point theory.

(Contemporary mathematics, ISSN 0271-4132; v. 14)

Based on courses given at the University of California, Berkeley, winter 1980, and at the University of California, Los Angeles, winter 1981.

Bibliography: p.

Includes index.

1. Fixed point theory. 2. Covering spaces (Topology) I. Title. II. Title: Nielsen fixed point theory. III. Series.

QA612.24.J5 1982 514'.2 82-20756

ISBN 0-8218-5014-8

TABLE OF CONTENTS

PREFACE

These notes are based on the topics courses given at the University of California, Berkeley, in Winter 1980 and at the University of California, Los Angeles, in Winter 1981. The subject is Nielsen fixed point theory which is becoming increasingly important in geometric topology and, potentially, has applications in analysis. The approach is via covering spaces. This approach is both natural and fruitful, but no reference in the English language has been easily available. The prerequisite is minimal: the classical covering space theory and homology theory for compact polyhedra.

The Introduction explains what Nielsen theory is about. Chapter I gives the basic notions of the theory, while Chapter II is devoted to computational methods. In Chapter III we broaden the scope and introduce the Nielsen type theory for periodic points. Chapter IV provides an exposition of the latest progress in the Nielsen theory for fiber maps. Another chapter in the original courses is now sketched as §I.6 because the material is easily available in the literature. The Historical Notes and Bibliography attached are by no means complete.

The author wishes to express his gratitude to Professor T. H. Kiang who introduced him into this subject years before and whose book [Kiang (1979)] has a great influence on the presentation here. He wishes to thank Professors R. Brown and H. Schirmer for their interest in the course and their encouragement and helpful comments. He is especially indebted to Professor Brown for his help with language and in proofreading. He also wants to thank the Department of Mathematics at UCLA for hospitality during his visit and for arranging the typing of these notes. He thanks Bob Neu for his skillful typing of this manuscript.

Peking, China -- Boju Jiang

Introduction

Let X be a space, and let $f : X \to X$ be a self-map. A fixed point of f is a solution of the equation $x = f(x)$. The set of all fixed points of f we will denote by $\mathrm{Fix}(f)$. Fixed point theory studies the nature of the fixed point set $\mathrm{Fix}(f)$ in relation to the space X and the map f, such as: existence (is $\mathrm{Fix}(f) \neq \emptyset$?); the number of fixed points $\#\mathrm{Fix}(f)$ (we will use the notation $\#S$ for the cardinality of a set S); the behavior under homotopy (how $\mathrm{Fix}(f)$ changes when f changes continuously); etc.

Fixed point theory started in the early days of topology, because of its close relationship with other branches of mathematics. Existence theorems are often proved by converting the problem into an appropriate fixed point problem. Examples are the existence of solutions for elliptic partial differential equations, and the existence of closed orbits in dynamical systems. In many problems, however, one is not satisfied with the mere existence of a solution. One wants to know the number, or at least a lower bound for the number of solutions. But the actual number of fixed points of a self-map can hardly be the subject of an interesting theory, since it can be altered by an arbitrarily small perturbation of the map. So, in topology, one proposes to determine the minimal number of fixed points in a homotopy class. This is what Nielsen fixed point theory is about. This is the theme of these notes.

Perhaps the best known fixed point theorem in topology is the Lefschetz fixed point theorem.

THEOREM (Lefschetz 1923; Hopf 1929) Let X be a compact polyhedron, and let $f : X \to X$ be a map. Define the Lefschetz number $L(f)$ of f to be

$$L(f) := \sum_q (-1)^q \operatorname{trace}(f_{q*} : H_q(X;Q) \to H_q(X;Q)) ,$$

where $H_*(X;Q)$ is the rational homology of X. If $L(f) \neq 0$, then every map homotopic to f has a fixed point.

The Lefschetz number is the total algebraic count of fixed points. It is a homotopy invariant and is easily computable. But it counts the fixed points "by multiplicity", just like what one does when one says an equation of degree n has n roots. So, the Lefschetz theorem, along with its special case, the Brouwer fixed point theorem, and its generalization, the widely used Leray-Schauder theorem in functional analysis, can tell existence only.

In contrast, the (chronologically) first result of Nielsen theory has set a beautiful example of a different type of theorem.

THEOREM (Nielsen-Brouwer 1921) Let $f : T^2 \to T^2$ be a self-map of the torus. Suppose the endomorphism induced by f on the fundamental group $\pi_1(T^2) \cong \mathbb{Z} \oplus \mathbb{Z}$ is represented by the 2×2 integral matrix A. Then the least number of fixed points in the homotopy class of f equals the absolute value of the determinant of $E - A$, where E is the identity matrix; in symbols,

$$\mathrm{Min}\{\#\mathrm{Fix}(g) \mid g \simeq f\} = |\det(E - A)| .$$

It can be shown that $\det(E - A)$ is exactly $L(f)$ on tori. This latter theorem says much more than the Lefschetz theorem specialized to the torus, since it gives a lower bound for the number of fixed points, or it confirms the existence of a homotopic map which is fixed point free. The proof was via the universal covering space $\mathbb{R}^2$ of the torus. From this instance evolved the central notions of Nielsen theory -- the fixed point classes and the Nielsen number.

Roughly speaking, Nielsen theory has two aspects. The geometric aspect concerns the comparison of the Nielsen number with the least number of fixed points in a homotopy class of self-maps. The algebraic aspect deals with the problem of computation for the Nielsen number. We choose to concentrate more on the latter aspect, partly because of the richness and difficulty of the theory, partly because of its importance to applications. As to the former aspect, we will confine ourselves to quoting the main results without proof, and recommend the books [Brown (1971)], Chapter VIII, and [Kiang (1979)], Chapter IV, for excellent expositions of earlier results, and the paper [Jiang (1980)] for the latest improvements and simplifications. We will also restrict our exposition to self-maps of compact polyhedra, since there seems to be no essential difficulty in extending further to compact ANRs or even to compact maps on noncompact ANRs by means of the method of domination (cf. [Brown (1969), (1971)] and [You]).

Nielsen theory is based on the theory of covering spaces. We will take this point of view consistently, as Nielsen himself did. An alternative way is to consider nonempty fixed point classes only, and use paths instead of covering spaces to define them. This is certainly more convenient for some geometric questions. But the covering space approach is theoretically more satisfactory, especially for computational problems, since the nonemptiness of certain fixed point classes is often the conclusion of the analysis, not the assumption.

Now let us introduce the basic idea of Nielsen theory by an elementary example.

PROPOSITION. Let $f : S^1 \to S^1$ be a self-map of the circle. Suppose the degree of f is d. Then the least number of fixed points in the homotopy class of f is $|1 - d|$.

Proof. Let S^1 be the unit circle on the complex plane, i.e. $S^1 = \{z \in \mathbb{C} \mid |z| = 1\}$. Let $p : \mathbb{R} \to S^1$ be the exponential map $p(\theta) = z = e^{i\theta}$. Then θ is the argument of z, which is a multi-valued function of z. For every $f : S^1 \to S^1$, one can always find "argument expressions" (or liftings) $\tilde{f} : \mathbb{R} \to \mathbb{R}$ such that $f(e^{i\theta}) = e^{i\tilde{f}(\theta)}$, in fact a whole series of them, differing from each other by integral multiples of 2π. For definiteness let us write $\tilde{f}_0$ for the argument expression with $\tilde{f}_0(0)$ lying in $[0, 2\pi)$, and write $\tilde{f}_k = \tilde{f}_0 + 2k\pi$. Since the degree of f is d, the functions $\tilde{f}_k$ are such that $\tilde{f}_k(\theta + 2\pi) = \tilde{f}_k(\theta) + 2d\pi$. For example, if $f(z) = -z^d$, then $\tilde{f}_k(\theta) = d\theta + (2k + 1)\pi$.

It is evident that if $z = e^{i\theta}$ is a fixed point of f, i.e. $z = f(z)$, then θ is a fixed point of some argument expression of f, i.e. $\theta = \tilde{f}_k(\theta)$ for some k. On the other hand, if θ is a fixed point of $\tilde{f}_k$, q is an integer, then $\theta + 2q\pi$ is a fixed point of $\tilde{f}_\ell$ iff $\ell - k = q(1 - d)$. This follows from the calculation $\tilde{f}_\ell(\theta + 2q\pi) = \tilde{f}_k(\theta + 2q\pi) + 2(\ell - k)\pi$ $= \tilde{f}_k(\theta) + 2qd\pi + 2(\ell - k)\pi = (\theta + 2q\pi) + 2\pi\{(\ell - k) - q(1 - d)\}$. Thus, if $\ell \not\equiv k \bmod (1 - d)$, then a fixed point of $\tilde{f}_k$ and a fixed point of $\tilde{f}_\ell$ can never correspond to the same fixed point of f, i.e. $p\,\mathrm{Fix}(\tilde{f}_k) \cap p\,\mathrm{Fix}(\tilde{f}_\ell) = \emptyset$.

So, the argument expressions fall into equivalence classes (called lifting classes) by the relation $\tilde{f}_k \sim \tilde{f}_\ell$ iff $k \equiv \ell \bmod (1 - d)$, and the fixed points of f split into $|1 - d|$ classes (called fixed point classes) of the form $p\,\mathrm{Fix}(\tilde{f}_k)$. That is, two fixed points are in the same class iff they come from fixed points of the same argument expression. Note that each fixed point class is by definition associated with a lifting class, so that the number of fixed point classes is $|1 - d|$ if $d \neq 1$, and is ∞ if $d = 1$. Also note that a fixed point class need not be nonempty.

Now, to prove that a map f of degree d has at least $|1 - d|$ fixed points, we only have to show that every fixed point class is nonempty, or equivalently, that every argument expression has a fixed point, if $d \neq 1$. In fact, for each k, by means of the equality $\tilde{f}_k(\theta + 2\pi) - \tilde{f}_k(\theta) = 2d\pi$, it is easily seen that the function $\theta - \tilde{f}_k(\theta)$ takes different signs when θ approaches $\pm\infty$, hence $\tilde{f}_k$ has at least one fixed point.

That $|1 - d|$ is indeed the least number of fixed points in the homotopy class is seen by checking the special map $f(z) = -z^d$. □

The following chapters can be considered as generalizations of this simplest example. See the table of contents and the introductory paragraph of every chapter.

CHAPTER 1

FIXED POINT CLASSES AND THE NIELSEN NUMBER

In this chapter we introduce the basic notions of Nielsen theory. The simple example of $S^1 \to S^1$ in the Introduction is generalized to self-maps of a polyhedron X, with the universal covering space $\tilde{X}$ of X playing the role of the exponential map $\mathbb{R} \to S^1$. The basic invariance theorems are in §§4-5. Section 3 is a brief introduction to the algebraic count of fixed points -- the fixed point index. We conclude this chapter by relating the Nielsen number to the least number of fixed points in a homotopy class, thus justifying the important position of the Nielsen number ir the fixed point theory.

1. LIFTING CLASSES AND FIXED POINT CLASSES. We always assume X to be a connected compact polyhedron. It is well known that X has a universal covering space. (Actually the material in §§1-2 makes sense for any X with a universal covering space.) References on covering spaces: [Massey], [Spanier].

Let $p : \tilde{X} \to X$ be the universal covering of X.

1.1 DEFINITION. A lifting of a map $X \xrightarrow{f} X$ is a map $\tilde{X} \xrightarrow{\tilde{f}} \tilde{X}$ such that $p \circ \tilde{f} = f \circ p$. A covering translation is a map $\tilde{X} \xrightarrow{\tilde{\gamma}} \tilde{X}$ such that $p \circ \gamma = p$, i.e. a lifting of the identity map.

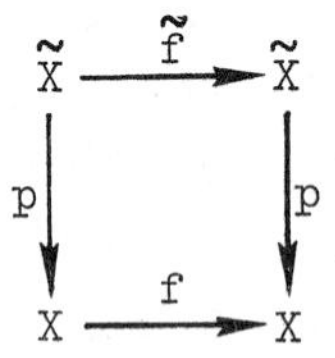

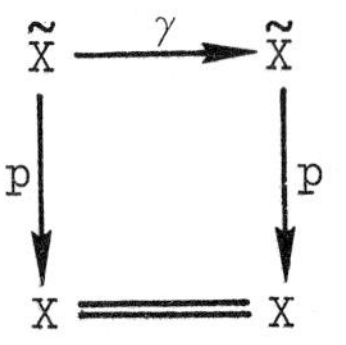

1.2 PROPOSITION. (i) For any $x_0 \in X$ and any $\tilde{x}_0, \tilde{x}_0' \in p^{-1}(x_0)$, there is a unique covering translation $\gamma : \tilde{X} \to \tilde{X}$ such that $\gamma(\tilde{x}_0) = \tilde{x}_0'$. The covering translations of $\tilde{X}$ form a group $\mathcal{D} = \mathcal{D}(\tilde{X},p)$ which is isomorphic to $\pi_1(X)$.

(ii) Let $f : X \to X$ be a map. For given $x_0 \in X$ and $x_1 = f(x_0)$, pick $\tilde{x}_0 \in p^{-1}(x_0)$ and $\tilde{x}_1 \in p^{-1}(x_1)$ arbitrarily. Then, there is a unique lifting

$\tilde{f}$ of f such that $\tilde{f}(\tilde{x}_0) = \tilde{x}_1$.

(iii) Suppose $\tilde{f}$ is a lifting of f, and $\alpha, \beta \in \mathcal{D}$. Then $\beta \circ \tilde{f} \circ \alpha$ is a lifting of f.

(iv) For any two liftings $\tilde{f}$ and $\tilde{f}'$ of f, there is a unique $\gamma \in \mathcal{D}$ such that $\tilde{f}' = \gamma \circ \tilde{f}$.

PROOF. (i) and (ii) are standard theorems in covering space theory. (iii) and (iv) follow from Definitions (i) and (ii). □

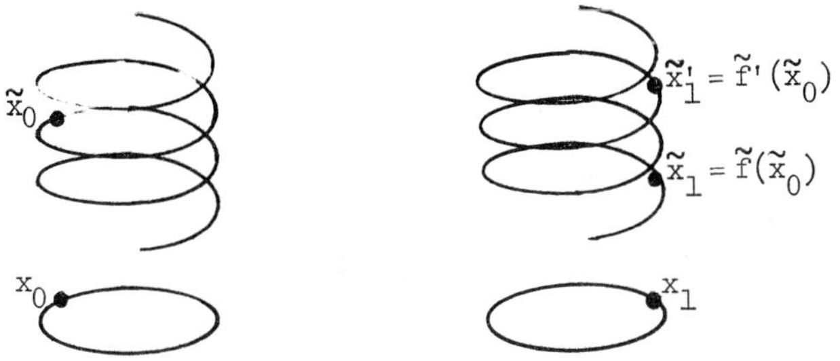

1.3 LEMMA. Suppose $\tilde{x} \in p^{-1}(x)$ is a fixed point of a lifting $\tilde{f}$ of f, and $\gamma \in \mathcal{D}$ is a covering translation on $\tilde{X}$. Then, a lifting $\tilde{f}'$ of f has $\gamma(\tilde{x}) \in p^{-1}(x)$ as a fixed point iff $\tilde{f}' = \gamma \circ \tilde{f} \circ \gamma^{-1}$.

PROOF. "If" is obvious: $\tilde{f}'(\gamma(\tilde{x})) = \gamma \circ \tilde{f} \circ \gamma^{-1}(\gamma(\tilde{x})) = \gamma \circ \tilde{f}(\tilde{x}) = \gamma(\tilde{x})$.

"Only if": Both $\tilde{f}'$ and $\gamma \circ \tilde{f} \circ \gamma^{-1}$ have $\gamma(\tilde{x})$ as a fixed point, so they agree at the point $\gamma(\tilde{x})$. By Proposition 1.2 (ii), they are the same lifting. □

1.4 DEFINITION. Two liftings $\tilde{f}$ and $\tilde{f}'$ of $f: X \to X$ are said to be <u>conjugate</u> if there exists $\gamma \in \mathcal{D}$ such that $\tilde{f}' = \gamma \circ \tilde{f} \circ \gamma^{-1}$. <u>Lifting classes</u> := equivalence classes by conjugacy. Notation:

$$[\tilde{f}] = \{\gamma \circ \tilde{f} \circ \gamma^{-1} \mid \gamma \in \mathcal{D}\}.$$

1.5 THEOREM. (i) $\mathrm{Fix}(f) = \bigcup_{\tilde{f}} p\,\mathrm{Fix}(\tilde{f})$.

(ii) $p\,\mathrm{Fix}(\tilde{f}) = p\,\mathrm{Fix}(\tilde{f}')$ if $[\tilde{f}] = [\tilde{f}']$.

(iii) $p\,\mathrm{Fix}(\tilde{f}) \cap p\,\mathrm{Fix}(\tilde{f}') = \emptyset$ if $[\tilde{f}] \neq [\tilde{f}']$.

PROOF. (i) If $x_0 \in \mathrm{Fix}(f)$, pick $\tilde{x}_0 \in p^{-1}(x_0)$. By Proposition 1.2 (ii) there exists $\tilde{f}$ such that $\tilde{f}(\tilde{x}_0) = \tilde{x}_0$. Hence $x_0 \in p\,\mathrm{Fix}(\tilde{f})$.

(ii) If $\tilde{f}' = \gamma \circ \tilde{f} \circ \gamma^{-1}$, then by Lemma 1.3, $\mathrm{Fix}(\tilde{f}') = \gamma\,\mathrm{Fix}(\tilde{f})$, so that $p\,\mathrm{Fix}(\tilde{f}') = p\,\mathrm{Fix}(\tilde{f})$.

(iii) If $x_0 \in p\,\mathrm{Fix}(\tilde{f}) \cap p\,\mathrm{Fix}(\tilde{f}')$, there are $\tilde{x}_0, \tilde{x}_0' \in p^{-1}(x_0)$ such that $\tilde{x}_0 \in \mathrm{Fix}(\tilde{f})$ and $\tilde{x}_0' \in \mathrm{Fix}(\tilde{f}')$. Suppose $\tilde{x}_0' = \gamma\tilde{x}_0$. By Lemma 1.3, $\tilde{f}' = \gamma \circ \tilde{f} \circ \gamma^{-1}$, hence $[\tilde{f}] = [\tilde{f}']$. □

1.6 DEFINITION. The subset $p\,\mathrm{Fix}(\tilde{f})$ of $\mathrm{Fix}(f)$ is called the <u>fixed point class</u> of f determined by the lifting class $[\tilde{f}]$.

1.7 THEOREM. The fixed point set $\mathrm{Fix}(f)$ splits into a disjoint union of fixed point classes. □

EXAMPLE. Lifting classes and fixed point classes of the identity map $\mathrm{id}_X : X \to X$.

A lifting class = a conjugacy class (in the usual sense) in $\mathcal{D}$.

$p\,\mathrm{Fix}(\mathrm{id}_{\tilde{X}}) = X$.

$p\,\mathrm{Fix}(\gamma) = \emptyset$ otherwise.

1.8 REMARK. A fixed point class is always considered to carry a label -- the lifting class determining it. Thus two empty fixed point classes are considered different if they are determined by different lifting classes.

1.9 DEFINITION. The number of lifting classes of f (and hence the number of fixed point classes, empty or not) is called the Reidemeister number of f, denoted $R(f)$. It is a positive integer or infinity.

EXAMPLE. $R(f) = 1$ if X is simply-connected.

Our definition of a fixed point class is via the universal covering space. It essentially says: Two fixed points of f are in the same class iff there is a lifting $\tilde{f}$ of f having fixed points above both of them. There is another way of saying this, which does not use covering space explicitly, hence is very useful in identifying fixed point classes.

1.10 THEOREM. Two fixed points x_0 and x_1 of $f : X \to X$ belong to the same fixed point class iff there is a path c from x_0 to x_1 such that $c \simeq f \circ c$ (homotopy rel endpoints).

PROOF. "Only if". Fixed points x_0 and x_1 are in the same class, then there exists a lifting $\tilde{f} : \tilde{X} \to \tilde{X}$ and points $\tilde{x}_0 \in p^{-1}(x_0)$ and $\tilde{x}_1 \in p^{-1}(x_1)$ such that $\tilde{f}(\tilde{x}_0) = \tilde{x}_0$ and $\tilde{f}(\tilde{x}_1) = \tilde{x}_1$.

Take a path $\tilde{c}$ in $\tilde{X}$ from $\tilde{x}_0$ to $\tilde{x}_1$. Since $\tilde{X}$ is simply-connected, $\tilde{c} \simeq \tilde{f} \circ \tilde{c}$. Projecting down to X, we have

$$c \simeq f \circ c$$

where $c = p \circ \tilde{c}$.

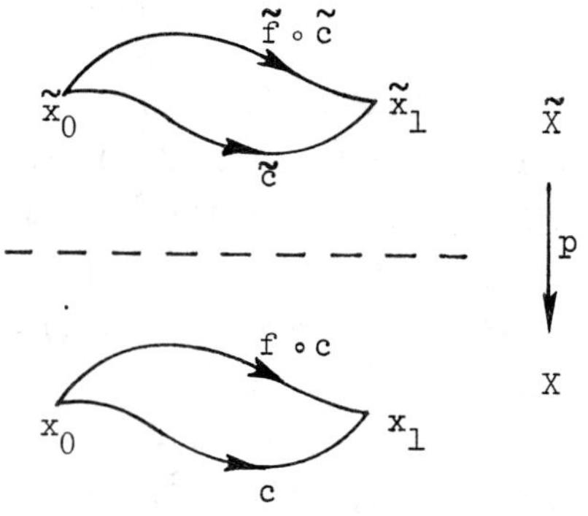

"If". Suppose $x_0 \in p\,\mathrm{Fix}(\tilde{f})$, $\tilde{x}_0 \in p^{-1}(x_0)$ and $\tilde{f}(\tilde{x}_0) = \tilde{x}_0$. We want to prove $x_1 \in p\,\mathrm{Fix}(\tilde{f})$, i.e. there exists $\tilde{x}_1 \in p^{-1}(x_1)$ such that $\tilde{f}(\tilde{x}_1) = \tilde{x}_1$.

Lift the path c from $\tilde{x}_0$ to get a path $\tilde{c}$ in $\tilde{X}$. Then $\tilde{f} \circ \tilde{c}$ projects to $f \circ c$, hence $\tilde{f} \circ \tilde{c}$ is the lift of $f \circ c$ from $\tilde{x}_0$.

Since $c \simeq f \circ c$, their lifts from the same starting point $\tilde{x}_0$ should have the same endpoint. Hence $\tilde{x}_1 = \tilde{f}(\tilde{x}_1)$, where $\tilde{x}_1$ is the other end of $\tilde{c}$. □

1.11 REMARK. Theorem 1.10 can be considered as an equivalent definition of a non-empty fixed point class. Its advantage: It works directly on X, hence is more convenient in geometric questions. Its disadvantage: It pays attention only to non-empty fixed point classes, hence is not satisfactory when considering the influence of homotopy on fixed point classes.

1.12 THEOREM. Every fixed point class of $f: X \to X$ is an open subset of $\mathrm{Fix}(f)$.

PROOF. Given a fixed point x_0 of f, we want to find a neighborhood U of x_0 such that any fixed point $x_1 \in U$ belongs to the same class.

Since X has a universal covering, X is locally path-connected and semilocally 1-connected. There is a neighborhood W of x_0 such that every loop in W at x_0 is trivial in X. There also is a path-connected neighborhood U of x_0 such that $U \subset W \cap f^{-1}(W)$.

Now, if $x_1 \in U \cap \mathrm{Fix}(f)$, take a path c in U from x_0 to x_1, then both c and $f \circ c$ are in W, hence $c \simeq f \circ c$. Thus x_0, x_1 are in the same class by Theorem 1.10. □

1.13 COROLLARY. Every map $f: X \to X$ has only finitely many non-empty fixed point classes, each a compact subset of X. □

1.14 COROLLARY. A continuum of fixed points lies in a single fixed point class. □

EXERCISES. 1. Let $f: S^1 \to S^1$ be of degree d. $R(f) = ?$

2. Let T^2 be the torus, $f: T^2 \to T^2$, the induced homomorphism on $H_1(T^2)$ given by an integral 2×2 matrix A. $R(f) = ?$

3. Discuss $R(f)$ for $f: \mathbb{R}P^2 \to \mathbb{R}P^2$.

2. THE INFLUENCE OF A HOMOTOPY. We use the following notation for a homotopy: $H = \{h_t\}_{t \in I}: f_0 \simeq f_1: X \to X$, or $H: X \times I \to X$.

Given a homotopy $H = \{h_t\}: f_0 \simeq f_1$, we want to see its influence on fixed point classes of f_0 and f_1.

2.1 DEFINITION. A homotopy $\tilde{H} = \{\tilde{h}_t\}: \tilde{X} \to \tilde{X}$ is called a lifting of the homotopy $H = \{h_t\}$, if $\tilde{h}_t$ is a lifting of h_t for every $t \in I$.

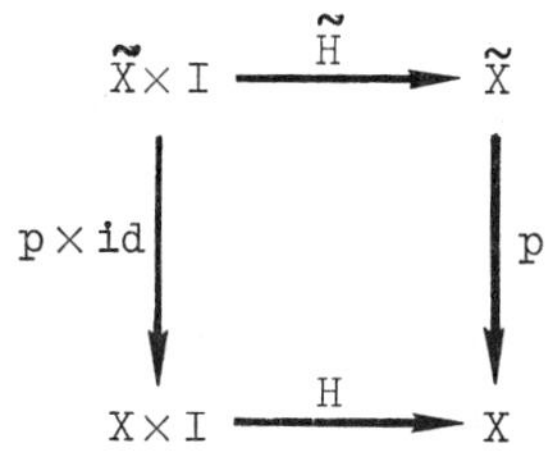

2.2 BASIC OBSERVATION. Given a homotopy $H : f_0 \simeq f_1$ and a lifting $\tilde{f}_0$ of f_0, there is a unique lifting $\tilde{H}$ of H such that $\tilde{h}_0 = \tilde{f}_0$, hence they determine a lifting $\tilde{f}_1$ of f_1. (Reason: Unique lifting property of covering spaces.) Thus H gives rise to a one-one correspondence from liftings of f_0 to liftings of f_1.

$$\tilde{f}_0 \overset{H}{\rightsquigarrow} \tilde{f}_1$$

$$\tilde{f}_0 \overset{H^{-1}}{\leftsquigarrow} \tilde{f}_1$$

This correspondence preserves the conjugacy relation:

$$\{\tilde{h}_t\} : \tilde{f}_0 \simeq \tilde{f}_1 \quad \text{implies} \quad \{\gamma \circ \tilde{h}_t \circ \gamma^{-1}\} : \gamma \circ \tilde{f}_0 \circ \gamma^{-1} \simeq \gamma \circ \tilde{f}_1 \circ \gamma^{-1} .$$

2.3 DEFINITION. Let $H : f_0 \simeq f_1$ be a homotopy and $\tilde{f}_i$ be a lifting of f_i, $i = 0,1$. We say that the lifting class $[\tilde{f}_0]$ (and the fixed point class $p\,\mathrm{Fix}(\tilde{f}_0)$ of f_0) corresponds to the lifting class $[\tilde{f}_1]$ (and the fixed point class $p\,\mathrm{Fix}(\tilde{f}_1)$ of f_1) via the homotopy H, if H has a lifting $\tilde{H} : \tilde{f}_0 \simeq \tilde{f}_1$.

2.4 THEOREM. If $f_0 \simeq f_1$, then there is a one-to-one correspondence between fixed point classes of f_0 and those of f_1. Hence $R(f)$ is a homotopy invariant. □

EXAMPLE 1. A non-empty fixed point class may disappear under a homotopy.

Consider maps $S^1 \to S^1$. The universal covering is $p : \mathbb{R} \to S^1$, given by $\theta \mapsto e^{i\theta}$.

Let $H = \{h_t : z \mapsto z e^{it\varepsilon}\}$, where $\varepsilon > 0$ is small. Take the lifting $\tilde{H} = \{\tilde{h}_t : \theta \mapsto \theta + t\varepsilon\}$. Then $p\,\mathrm{Fix}(\tilde{h}_0) = S^1$ but $p\,\mathrm{Fix}(\tilde{h}_1) = \emptyset$.

EXAMPLE 2. The correspondence may depend on the homotopy H.

Consider maps $S^1 \to S^1$. Let $f_0 = f_1 : z \to z^{-2}$. Consider two homotopies $H' = \{h'_t : z \mapsto z^{-2}\} : f_0 \simeq f_1$ and $H = \{h_t : z \to z^{-2} e^{2\pi t i}\} : f_0 \simeq f_1$.

Take $\tilde{f}_0 : \theta \mapsto -2\theta$, then H' and H lift to $\tilde{H}' = \{\tilde{h}'_t : \theta \mapsto -2\theta\}$ and $\tilde{H} = \{\tilde{h}_t : \theta \mapsto -2\theta + 2\pi t\}$ respectively. So $\tilde{f}_0$ corresponds to $\tilde{f}'_1 : \theta \mapsto -2\theta$ via H', but corresponds to $\tilde{f}_1 : \theta \mapsto -2\theta + 2\pi$ via H. Thus $p\,\mathrm{Fix}(\tilde{f}_0) = \{1\}$ corresponds to $p\,\mathrm{Fix}(\tilde{f}'_1) = \{1\}$ via H', but corresponds to $p\,\mathrm{Fix}(\tilde{f}_1) = \{e^{2\pi i/3}\}$ via H.

We now turn to another view of the above correspondence.

2.5 DEFINITION. Every homotopy $H : X \times I \to X$ gives rise to a level-preserving map $\mathbb{H} : X \times I \to X \times I$ in an obvious way:

$$\mathbb{H}(x,t) = (H(x,t),t) = (h_t(x),t) .$$

The map $\mathbb{H}$ will be called the <u>fat homotopy</u> of H, and h_t will be called the t-<u>slice</u> of $\mathbb{H}$. Similarly, for a subset $A \subset X \times I$, the subset $A_t := \{x \in X \mid (x,t) \in A\} \subset X$ will be called the t-<u>slice</u> of A.

The advantage of considering $\mathbb{H}$ is that it is a self-map of $X \times I$, so we may talk about its liftings and fixed point classes. Note that the universal covering of $X \times I$ is $p \times \mathrm{id}: \tilde{X} \times I \to X \times I$.

2.6 OBSERVATION. A lifting $\tilde{H}$ of $H \longleftrightarrow$ A lifting $\tilde{\mathbb{H}}$ of $\mathbb{H}$.
$\tilde{H}$ is a lifting of $H \Longleftrightarrow \tilde{\mathbb{H}}$ is a lifting of $\mathbb{H}$.
$\tilde{f}_0$ and $\tilde{f}_1$ correspond via $H \Longleftrightarrow \tilde{f}_0$ and $\tilde{f}_1$ are slices of an $\tilde{\mathbb{H}}$.

2.7 THEOREM. Let $H: f_0 \simeq f_1$ be a homotopy, $\mathbb{H}$ be its fat homotopy. Let $\tilde{f}_0, \tilde{f}_1$ be liftings of f_0, f_1 respectively, and let $\mathbb{F}_0 = p\,\mathrm{Fix}(\tilde{f}_0)$ and $\mathbb{F}_1 = p\,\mathrm{Fix}(\tilde{f}_1)$ be fixed point classes of f_0, f_1 respectively. Then $[\tilde{f}_0]$ corresponds to $[\tilde{f}_1]$ via H iff they are, respectively, the 0- and 1-slices of a single lifting class of $\mathbb{H}$; and $\mathbb{F}_0$ corresponds to $\mathbb{F}_1$ via H iff they are respectively the 0- and 1-slices of a single fixed point class of $\mathbb{H}$. □

This theorem is nothing but a restatement of the basic definition 2.3 in the language of fat homotopies. But it does reduce the identification of a correspondence via homotopy to the identification of a fixed point class. Thus, by 1.14 we have

2.8 COROLLARY. Let $H: f_0 \simeq f_1$ be a homotopy. Let $x_0 \in \mathrm{Fix}(f_0)$ and $x_1 \in \mathrm{Fix}(f_1)$. If $(x_0,0)$ and $(x_1,1)$ are connected by a continuum of fixed points of the fat homotopy $\mathbb{H}$, then the class of x_0 corresponds to the class of x_1 via H. □

Combining Theorem 2.7 with Theorem 1.10, we get

2.9 THEOREM. Let $H = \{h_t\}: f_0 \simeq f_1: X \to X$ be a homotopy, $x_0 \in \mathrm{Fix}(f_0)$ and $x_1 \in \mathrm{Fix}(f_1)$. Suppose x_0 belongs to a fixed point class $\mathbb{F}_0$ of f_0, and x_1 belongs to a fixed point class $\mathbb{F}_1$ of f_1. Then $\mathbb{F}_0$ corresponds to $\mathbb{F}_1$ via H iff there is a path $c = \{x_t\}_{t \in I}$ in X from x_0 to x_1 such that $\{h_t(x_t)\} \simeq \{x_t\}$ with endpoints fixed.

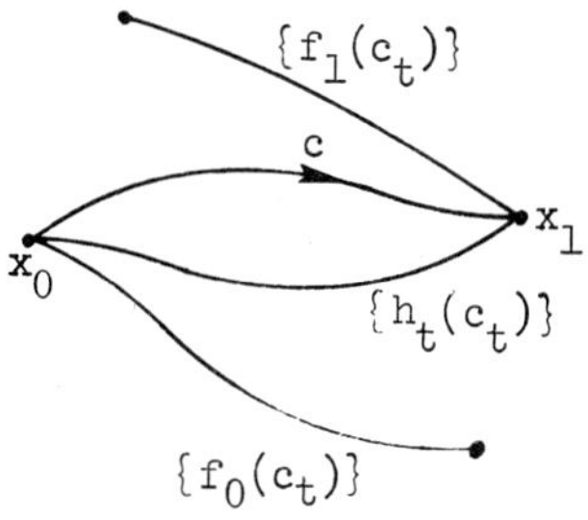

PROOF. $\mathbb{F}_0 \overset{H}{\dashrightarrow} \mathbb{F}_1 \overset{2.7}{\Longleftrightarrow} (x_0,0)$ and $(x_1,1)$ lie in the same fixed point class of $\mathbb{H} \overset{1.10}{\Longleftrightarrow}$ there is a path $\{(x_t,s_t)\}$ in $X \times I$ from $(x_0,0)$ to $(x_1,1)$ such that $\{\mathbb{H}(x_t,s_t)\} = \{(h_t(x_t),s_t)\} \simeq \{(x_t,s_t)\}$, which is obviously equivalent to $\{h_t(x_t)\} \simeq \{x_t\}$. □

2.10 REMARK. This theorem can be considered as an equivalent definition of correspondence via a homotopy, for <u>non-empty</u> fixed point classes. Compare Remark 1.11.

There is still another geometric characterization of correspondence via a homotopy.

2.11 DEFINITION. A <u>deformation</u> of a homotopy $H : f_0 \simeq f_1 : X \to X$ into another $H' : f_0 \simeq f_1$ is a continuous family $\{H_u : f_0 \simeq f_1\}_{u \in I}$ with $H_0 = H$ and $H_1 = H'$.

2.12 OBSERVATION. If two homotopies $H, H' : f_0 \simeq f_1$ are deformable into each other, then they give rise to the same correspondence from the fixed point classes of f_0 to the fixed point classes of f_1. In fact, if H lifts to $\tilde{H} : \tilde{f}_0 \simeq \tilde{f}_1 : \tilde{X} \to \tilde{X}$, then the deformation $\{H_u\}_{u \in I}$ lifts to a deformation $\{\tilde{H}_u : \tilde{f}_0 \simeq \tilde{f}_1\}_{u \in I}$.

2.13 THEOREM. Let $H : f_0 \simeq f_1 : X \to X$ be a homotopy, $x_0 \in \mathrm{Fix}(f_0)$ and $x_1 \in \mathrm{Fix}(f_1)$. Then, the class of x_0 corresponds to the class of x_1 via H iff H is deformable into a homotopy $H' = \{h_t'\}_{t \in I}$ such that there is a path $c = \{x_t\}_{t \in I}$ from x_0 to x_1 with $x_t \in \mathrm{Fix}(h_t')$ for all $t \in I$.

PROOF. The "if" part follows easily from Observation 2.12 and Theorem 2.9. It remains to prove the "only if" part.

By Theorem 2.9, there is a path $c = \{x_t\}$ from x_0 to x_1 such that $\{h_t(x_t)\} \simeq \{x_t\}$, i.e. there is a $D : I \times I \to X$ with $D(0,s) = x_0$, $D(1,s) = x_1$, $D(t,0) = h_t(x_t)$, and $D(t,1) = x_t$ for all $t, s \in I$.

The polyhedron X is uniformly locally contractible (cf. [Brown (1971)], p. 39), i.e. there exists a map $\gamma : W \times I \to X$, where $W = \{(x,x') \in X \times X \mid d(x,x') < \delta\}$ for some $\delta > 0$, such that $\gamma(x,x',0) = x$, $\gamma(x,x',1) = x'$, and $\gamma(x,x,t) = x$ for all $x, x' \in X$, $t \in I$.

Let $G : \{(x,t) \mid (x,x_t) \in W\} \times I \to X$ be defined by

$$G(x,t,s) = \begin{cases} h_t(\gamma(x,x_t,2s)) & \text{if } s \le \frac{1}{2}, \\ D(t,2s-1) & \text{if } s \ge \frac{1}{2}. \end{cases}$$

It is obviously continuous. Let $\theta(t) = \delta t(1-t)$. Define a deformation $\{H_u\}_{u \in I} : X \times I \to X$ by

$$H_u(x,t) = \begin{cases} H(x,t) & \text{if } d(x,x_t) \ge \theta(t), \\ G(x,t,u - ud(x,x_t)/\theta(t)), & \text{if } d(x,x_t) \le \theta(t) > 0. \end{cases}$$

Note that $\{H_u\}$ is well-defined. The continuity is obvious except at the points with $d(x,x_t) = \theta(t) = 0$, i.e. at points with $x = x_0$, $t = 0$ or $x = x_1$, $t = 1$. The continuity at these points follows from the fact that $G(x_0,0,s) = x_0$ and $G(x_1,1,s) = x_1$ for all $s \in I$. It is easy to check that $H_u : f_0 \simeq f_1$ and $H_0 = H$, and that $H_1 = H'$ satisfies $H'(x_t,t) = x_t$, i.e. $x_t \in \mathrm{Fix}(h_t')$. □

EXERCISE. Let $H, H' : f_0 \simeq f_1 : X \to X$ be two homotopies connecting f_0 and f_1. Show that: H and H' give the same correspondence from liftings of f_0 to liftings of f_1 iff for some (hence every) point $x \in X$ the paths $\{h_t(x)\}_{t\in I}$ and $\{h'_t(x)\}_{t\in I}$ are homotopic with endpoints fixed.

3. THE FIXED POINT INDEX. The fixed point index is an indispensable tool of fixed point theory. It provides an algebraic count of fixed points in an open set. There are many different approaches to the fixed point index, all turn out to be equivalent, hence an axiomatic approach has emerged and existence and uniqueness proved. Instead of giving a self-contained treatment, we will introduce a naive, step-by-step construction of this index, and list (without proof) the most useful properties. The serious reader may consult the books [Alexandroff-Hopf], [Brown(1971)] and [Dold (1972)].

(A) THE INDEX OF AN ISOLATED FIXED POINT IN $\mathbb{R}^n$. A reasonable algebraic count of fixed points should be a generalization of the notion of multiplicity for zeros of a complex analytic function.

Suppose $\mathbb{R}^n \supset U \xrightarrow{f} \mathbb{R}^n$, and $a \in U$ is an isolated fixed point of f. Pick a sphere S^{n-1}_a centered at a, small enough to exclude other fixed points. On S^{n-1}_a, the vector $x - f(x) \neq 0$, so a direction field $\varphi : S^{n-1}_a \to S^{n-1}$, $\varphi(x) = \dfrac{x - f(x)}{|x - f(x)|}$, is defined.

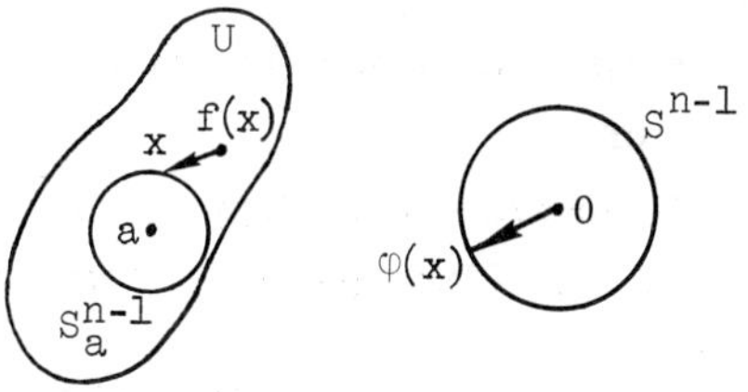

3.1 DEFINITION. index(f,a) = degree of φ.

This definition doesn't depend on the radius of S^{n-1}_a.

EXAMPLE 1. $n = 1$. The local picture in $\mathbb{R}^2 = \mathbb{R} \times \mathbb{R}$ of an intersection of the diagonal with the graph of f.

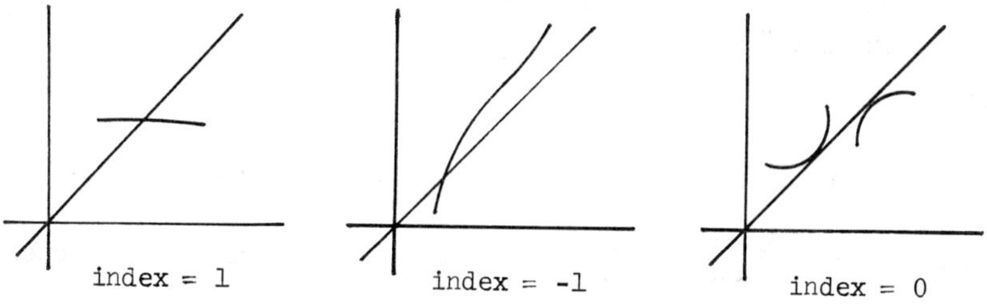

EXAMPLE 2. $n = 2$. Suppose f has a complex analytic expression $z \mapsto f(z)$ Then a fixed point of f is nothing but a zero of the function $z - f(z)$. Suppose z_0 is a fixed point of f. It follows from the theory of analytic functions that the multiplicity of the zero z_0 = the variance (counted by multiples of 2π) of $\arg(z - f(z))$ when z moves around z_0 once = deg φ.

So $\operatorname{index}(f,z_0)$ = multiplicity of the zero z_0 of the analytic function $z - f(z)$.

This example shows that when $n = 2$, in contrast to the previous example, the index of an isolated fixed point can be arbitrarily large.

To show a negative index, let $f(z)$ be given (locally) as $f(z) = z - \overline{F(z)}$, where $F(z)$ is analytic. Then $z - f(z) = \overline{F(z)}$, so we get a negative degree, hence a negative index.

3.2 PROPERTIES. (1) Constant map. Suppose $\mathbb{R}^n \supset U \xrightarrow{f} a \in U \subset \mathbb{R}^n$ is a constant map. Then

$$\operatorname{index}(f,a) = 1 . \qquad \square$$

(2) Differentiable map. Suppose $\mathbb{R}^n \supset U \xrightarrow{f} \mathbb{R}^n$, and $a \in U$. Suppose f is differentiable at a with Jacobian $A = (\frac{\partial f}{\partial x})_a$, and $\det(E - A) \neq 0$ where E is the identity matrix. Then a is an isolated fixed point and

$$\begin{aligned}\operatorname{index}(f,a) &= \operatorname{sgn}\det(E - A) . \\ &= (-1)^k\end{aligned}$$

where k is the number (counted with multiplicity) of real eigenvalues greater than 1 of A. $\square$

(3) Restriction. Suppose $\mathbb{R}^n \supset U \xrightarrow{f} \mathbb{R}^m \subset \mathbb{R}^n$, $m < n$, and $a \in U \cap \mathbb{R}^m$ is an isolated fixed point. If we consider $\mathbb{R}^m \supset U \cap \mathbb{R}^m \xrightarrow{f} \mathbb{R}^m$, then

$$\operatorname{index}(f,a) = \operatorname{index}(f \mid U \cap \mathbb{R}^m, a) .$$

PROOF. It suffices to prove the case $m = n - 1$. Consider the degree of the suspension of a direction field. $\square$

(4) Multiplicativity. Suppose $\mathbb{R}^n \supset U \xrightarrow{f} \mathbb{R}^n$ and $\mathbb{R}^m \supset V \xrightarrow{g} \mathbb{R}^m$, with $a \in U$ and $b \in V$. They give rise to the product $\mathbb{R}^{n+m} \supset U \times V \xrightarrow{f\times g} \mathbb{R}^{n+m}$, and $a \times b \in U \times V$.

Then

$$\operatorname{index}(f \times g, a \times b) = \operatorname{index}(f,a) \cdot \operatorname{index}(g,b) .$$

PROOF. We have $S^{n+m-1}_{a\times b} = S^{n-1}_a \circ S^{m-1}_b$, where $\circ$ is the join operation, and

$$\varphi \mid S^{n+m-1}_{a\times b} = (\varphi \mid S^{n-1}_a) \circ (\varphi \mid S^{m-1}_b) .$$

Hence $\deg(\varphi \mid S^{n+m-1}_{a\times b}) = \deg(\varphi \mid S^{n-1}_a) \cdot \deg(\varphi \mid S^{m-1}_b)$. $\square$

5. Removability. Suppose $\mathbb{R}^n \supset U \xrightarrow{f} \mathbb{R}^n$, and $a \in U$ is an isolated fixed point with $\operatorname{index}(f,a) = 0$. Then this fixed point is removable.

PROOF. Since $\text{index}(f,a) = \deg(\varphi : S_a^{n-1} \to S^{n-1}) = 0$, the direction field $\varphi : S_a^{n-1} \to S^{n-1}$ is extendable to a map $\hat{\varphi} : D_a^n \to S^{n-1}$ on the disk D_a^n, by a theorem of Hopf. Hence we may redefine f on D_a^n to get rid of the fixed point: Let $\lambda : S_a^{n-1} \to \mathbb{R}$ be $\lambda(x) = \|x - f(x)\|$, with $\lambda(x) > 0$ on S_a^{n-1}. Extend λ to $\hat{\lambda} : D_a^n \to \mathbb{R}$ with $\hat{\lambda}(x) > 0$ on D_a^n. Then set

$$\hat{f}(x) = \begin{cases} f(x) & \text{if } x \notin D_a^n , \\ x - \hat{\lambda}(x)\hat{\varphi}(x) & \text{if } x \in D_a^n . \end{cases}$$

□

(B) FIXED POINT INDEX IN $\mathbb{R}^n$.

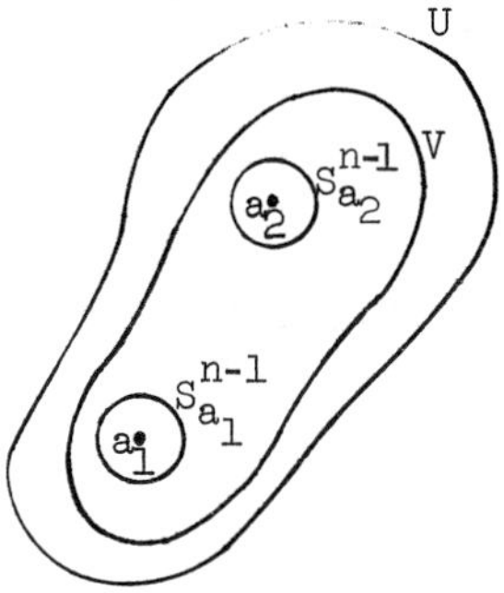

First, imagine we have $\mathbb{R}^n \supset U \xrightarrow{f} \mathbb{R}^n$ such that all fixed points $\{a_j\}$ of f are isolated. Pick a small $S_{a_j}^{n-1}$ for each a_j and define the direction field $\varphi : U - \text{Fix}(f) \to S^{n-1}$ as before. Suppose $\varphi S_{a_j}^{n-1} = i_j S^{n-1}$ in the homological sense (that is, $i_j = \deg(\varphi \mid S_{a_j}^{n-1})$), then $i_j = \text{index}(f,a_j)$.

It would be reasonable to define $\text{index}(f,U) = \Sigma_j\, i_j = \Sigma_j\, \text{index}(f,a_j)$. But this number has another interpretation. Take an open subset V of U such that $\bar{V} \subset U$, V contains all the fixed points of f, and $\bar{V}$ is a smooth n-manifold in U. Now φ is defined on ∂V, so we can talk about the integer i such that $\varphi(\partial V) = i\, S^{n-1}$ (also in the homological sense). We claim that this i is the sum of those i_j's. In fact, since φ is defined on $\bar{V} - \Sigma_j D_{a_j}^n$, we have

$$\varphi(\partial V - \Sigma_j S_{a_j}^{n-1}) = \varphi(\partial(\bar{V} - \Sigma_j D_{a_j}^n)) = \partial\varphi(\bar{V} - \Sigma_j D_{a_j}^n) = 0 ,$$

hence $i\, S^{n-1} - (\Sigma_j\, i_j)\, S^{n-1} = 0$, which gives $i = \Sigma_j\, i_j$. This analysis motivates the following definition which applies to non-isolated fixed points as well.

3.3 DEFINITION. Suppose $\mathbb{R}^n \supset U \xrightarrow{f} \mathbb{R}^n$, and $\text{Fix}(f)$ is compact. Take any open set $V \subset U$ such that $\text{Fix}(f) \subset V \subset \bar{V} \subset U$ and $\bar{V}$ is a smooth (or triangulable) n-manifold, then $\varphi(\partial V) = i\, S^{n-1}$ in the homological sense for some $i \in \mathbb{Z}$. This i is independent of the choice of V, and is defined to be the <u>fixed point index</u> of f on U, denoted $\text{index}(f,U)$.

Another way of defining $\text{index}(f,U)$ is by approximation, namely, approximate f by a smooth map with only generic fixed points (isolated fixed points satisfying the condition in 3.2 (2)), and add up their indices.

(C) FIXED POINT INDEX FOR POLYHEDRA. Every compact polyhedron can be embedded in some Euclidean space as a neighborhood retract.

Suppose now we are given a compact polyhedron X, and a map $X \supset U \xrightarrow{f} X$. X can be imbedded in $\mathbb{R}^N$ with inclusion $X \xrightarrow{i} \mathbb{R}^N$. And there is a neighborhood W of $i(X)$ in $\mathbb{R}^N$ and a retraction $W \xrightarrow{r} X$ such that $r \circ i = \mathrm{id}_X$. We have a diagram

$$\begin{array}{ccccc} X & \supset & U & \xrightarrow{f} & X \\ \uparrow r & & \uparrow r & & \downarrow i \\ \mathbb{R}^N \supset W & \supset & r^{-1}(U) & \xrightarrow{i \circ f \circ r} & \mathbb{R}^N \end{array}$$

3.4 DEFINITION. When $\mathrm{Fix}(f)$ is compact, define the <u>fixed point index</u> to be

$$\mathrm{index}(f,U) := \mathrm{index}(i \circ f \circ r, r^{-1}(U)) ,$$

the latter being the index in $\mathbb{R}^N$. It is independent of the choice of N, W, i and r.

All the facts we need in these Notes about the fixed point index are listed below.

3.5 BASIC FACTS ABOUT THE FIXED POINT INDEX. Let X be a compact polyhedron, $U \subset X$ an open subset, $f : U \to X$ a map such that $\mathrm{Fix}(f)$ (the fixed point set of f in U) is compact. For every such $X \supset U \xrightarrow{f} X$, an integer $\mathrm{index}(f,U)$ is defined, called the <u>fixed point index</u> of f in U, having the following properties:

(i) Existence of fixed points. If $\mathrm{index}(f,U) \neq 0$, then f has at least one fixed point in U.

(ii) Homotopy invariance. If $H = \{h_t\} : f_0 \simeq f_1 : U \to X$ is a homotopy such that $\bigcup_{t\in I} \mathrm{Fix}(h_t)$ is compact, then

$$\mathrm{index}(f_0,U) = \mathrm{index}(f_1,U) .$$

(iii) Additivity. Suppose $U_1,\ldots,U_s$ are disjoint open subsets of U, and f has no fixed point on $U - \bigcup_{j=1}^{s} U_j$. If $\mathrm{index}(f,U)$ is defined, then $\mathrm{index}(f,U_j)$, $j = 1,\ldots,s$ are all defined and

$$\mathrm{index}(f,U) = \Sigma_{j=1}^{s} \mathrm{index}(f,U_j) .$$

Here $\mathrm{index}(f,U_j)$ is shorthand for $\mathrm{index}(f \mid U_j, U_j)$.

(iv) Multiplicativity. Suppose $X \supset U \xrightarrow{f} X$ and $Y \supset V \xrightarrow{g} Y$. Consider the product $X \times Y \supset U \times V \xrightarrow{f \times g} X \times Y$. If $\mathrm{index}(f,U)$ and $\mathrm{index}(g,V)$ are defined, so is $\mathrm{index}(f \times g, U \times V)$, and

$$\mathrm{index}(f \times g, U \times V) = \mathrm{index}(f,U) \cdot \mathrm{index}(g,V) .$$

(v) Commutativity. Let X, Y be polyhedra, $U \subset X$ and $V \subset Y$ be open subsets, $f : U \to Y$ and $g : V \to X$ be maps. Then, the composite maps $X \supset f^{-1}(V) \xrightarrow{g \circ f} X$ and $Y \supset g^{-1}(U) \xrightarrow{f \circ g} Y$ are defined, and f, g restrict to a pair of homeomorphisms between the fixed point sets $\mathrm{Fix}(g \circ f) \underset{g}{\overset{f}{\rightleftarrows}} \mathrm{Fix}(f \circ g)$. If $\mathrm{index}(g \circ f, f^{-1}(V))$ is defined, so is $\mathrm{index}(f \circ g, g^{-1}(U))$, and

$$\mathrm{index}(g \circ f, f^{-1}(V)) = \mathrm{index}(f \circ g, g^{-1}(U)) .$$

(vi). Normalization. If $f : X \to X$, then

$$\mathrm{index}(f,X) = L(f) := \Sigma_q \ (-1)^q \mathrm{trace}(f_{q*} : H_q(X;Q) \to H_q(X;Q)) ,$$

where Q stands for the field of rational numbers. The number $L(f)$ is called the <u>Lefschetz number</u> of f.

(vii). Removability Condition. Suppose $X \supset U \xrightarrow{f} X$ has only one fixed point x^*, with $\mathrm{index}(f,U) = 0$, and x^* has a Euclidean neighborhood. Then, given any neighborhood V of x^*, there is a map $g \simeq f : U \to X$ such that $g = f$ on $U - V$ and g has no fixed point on U.

The properties (ii), (iii) for $s = 2$, (v) and (vi) may be used as axioms for the fixed point index, as in [Brown (1971)].

3.6 COROLLARY. Topological invariance. Suppose $h : X \to Y$ is a homeomorphism and the diagram

$$\begin{array}{ccccc} X & \supset & U & \xrightarrow{\ f\ } & X \\ {\scriptstyle h}\downarrow{\scriptstyle \cong} & & {\scriptstyle h}\downarrow{\scriptstyle \cong} & & {\scriptstyle h}\downarrow{\scriptstyle \cong} \\ Y & \supset & V & \xrightarrow{\ g\ } & Y \end{array}$$

commutes. Then $\mathrm{index}(f,U) = \mathrm{index}(g,V)$.

(Note that when we defined the index of an isolated fixed point, we used the Euclidean structure of $\mathbb{R}^n$, so Definition 3.1 is not topologically invariant in itself.)

PROOF. Using commutativity on $U \xrightarrow{h\circ f} Y$ and $V \xrightarrow{h^{-1}} X$, we get

$$\mathrm{index}(f,f^{-1}(U)\cap U) = \mathrm{index}(g,V) .$$

By additivity, $\mathrm{index}(f,U) = \mathrm{index}(f,f^{-1}(U)\cap U)$. □

3.7 COROLLARY. Restriction. Suppose $X \supset U \xrightarrow{f} X$ maps U into $Y \subset X$. Then $\mathrm{index}(f,U) = \mathrm{index}(f\,|\,U\cap Y, U\cap Y)$.

PROOF. Applying commutativity to $U \xrightarrow{f} Y$ and $U\cap Y \xrightarrow{i} X$, we get

$$\mathrm{index}(f,f^{-1}(U)\cap U) = \mathrm{index}(f\,|\,U\cap Y, U\cap Y) .$$

And $\mathrm{index}(f,f^{-1}(U)\cap U) = \mathrm{index}(f,U)$ by additivity. □

(D) THE INDEX OF AN ISOLATED SET OF FIXED POINTS. Let X be a compact polyhedron, $U \subset X$ an open set, $f: U \to X$ a map such that $\mathrm{Fix}(f)$ is compact. A set of fixed points $S \subset \mathrm{Fix}(f)$ is called an <u>isolated set of fixed points</u> if S is compact and open in $\mathrm{Fix}(f)$, i.e. if both S and $\mathrm{Fix}(f) - S$ are compact.

3.8 DEFINITION. The <u>index</u> of an isolated set S of fixed points, denoted $\mathrm{index}(f,S)$, is defined as follows. Pick a neighborhood $W \subset U$ of S isolating S from other fixed points, i.e. such that $S = W \cap \mathrm{Fix}(f)$. Define

$$\mathrm{index}(f,S) := \mathrm{index}(f,W) \qquad (\text{strictly speaking, } \mathrm{index}(f\,|\,W,W))$$

It is well defined: If W' is another such neighborhood, by additivity we have $\mathrm{index}(f,W) = \mathrm{index}(f,W\cap W') = \mathrm{index}(f,W')$.

Obviously each isolated fixed point is by itself an isolated set of fixed points, so we can talk about the index of an isolated fixed point. It coincides with the index in Definition 3.1 if that point has a Euclidean neighborhood.

A corollary of the commutativity property 3.5 (v) is:

3.9 COROLLARY. Let the notation be as in 3.5 (v). If S is an isolated set of fixed points of $g\circ f$, then $f(S)$ is an isolated set of fixed points of $f\circ g$, and

$$\mathrm{index}(g\circ f,S) = \mathrm{index}(f\circ g,f(S)) .$$

PROOF. Pick an isolating neighborhood $W \subset f^{-1}(V)$ for S. Then $g^{-1}(W)$ is obviously an isolating neighborhood for $f(S)$. Applying 3.5 (v) to the maps

$$X \supset W \xrightarrow{f|W} Y \qquad \text{and} \qquad Y \supset V \xrightarrow{g} X ,$$

we get

$$\mathrm{index}(g \circ f, W) = \mathrm{index}(f \circ g, g^{-1}(W)) .$$

This is just what we want, by Definition 3.8. □

A corollary of the homotopy invariance 3.5 (ii) is:

3.10 COROLLARY. Let the notation be as in 3.5 (ii). Let $\mathbb{H} : U \times I \to X \times I$ be the (level-preserving) fat homotopy given by H. Let S be an isolated set of fixed points of $\mathbb{H}$. Then, for any $t \in I$, the t-slice S_t is an isolated set of fixed points for h_t, and

$$\mathrm{index}(h_t, S_t) = \mathrm{index}(\mathbb{H}, S) .$$

PROOF. Take an isolating neighborhood $W \subset U \times I$ for S. Then the t-slice W_t is an isolating neighborhood of S_t. According to Definition 3.8, we should prove $\mathrm{index}(h_t, W_t) = \mathrm{index}(\mathbb{H}, W)$.

The idea is to squeeze $\mathbb{H}$ into a "thin" map resembling h_t. Define a homotopy $\{G_u\} : X \times I \to X \times I$ by

$$G_u(x,s) = (h_{(1-u)s+ut}(x), (1-u)s + ut) .$$

Evidently $G_0 = \mathbb{H}$, and $\bigcup_u \mathrm{Fix}(G_u) \cap W = \mathrm{Fix}(\mathbb{H}) \cap W$ is compact, hence by 3.5 (ii) we have $\mathrm{index}(\mathbb{H}, W) = \mathrm{index}(G_1, W)$. Now apply 3.5 (v) to the map

$$X \times I \supset W \xrightarrow{H} X \quad \text{and} \quad X \xrightarrow{i_t} X \times I ,$$

where $i_t(x) = (x,t)$. We get

$$\begin{aligned} \mathrm{index}(h_t, W_t) &= \mathrm{index}(H \circ i_t, i_t^{-1}(W)) \\ &= \mathrm{index}(i_t \circ H, W) \\ &= \mathrm{index}(G_1, W) . \end{aligned}$$

Hence $\mathrm{index}(\mathbb{H}, W) = \mathrm{index}(h_t, W_t)$. □

4. THE NIELSEN NUMBER AND ITS HOMOTOPY INVARIANCE. Let $f : X \to X$ be a self-map of a connected compact polyhedron, and let $\mathbb{F}$ be a fixed point class of f.

According to Theorem 1.12, $\mathbb{F}$ is an isolated set of fixed points. So, $\mathrm{index}(f, \mathbb{F})$ is defined as in Definition 3.8.

4.1 DEFINITION. $\mathbb{F}$ is <u>essential</u> if $\mathrm{index}(f, \mathbb{F}) \neq 0$, <u>inessential</u> if $\mathrm{index}(f, \mathbb{F}) = 0$.

4.2 DEFINITION. The number of essential fixed point classes of f is called the <u>Nielsen number</u> of f, denoted $N(f)$.

The next two theorems follow directly from the definitions and the properties 3.5 of the fixed point index.

4.3 THEOREM. (i) $N(f) \leq R(f)$.

(ii) Each essential fixed point class is non-empty.

(iii) $N(f)$ is a non-negative integer, $0 \leq N(f) < \infty$.

(iv) $N(f) \leq \#\mathrm{Fix}(f)$. □

4.4 THEOREM. The sum of the indices of all (essential) fixed point classes of f equals $L(f)$. In symbols,

$$\sum_j \operatorname{index}(f, \mathbb{F}_j) = L(f) .$$

Hence $L(f) \neq 0$ implies $N(f) \geq 1$. □

EXAMPLE 1. $N(\mathrm{id}_X) = \begin{cases} 1 & \text{if } \chi(X) \neq 0 , \\ 0 & \text{if } \chi(X) = 0 . \end{cases}$

EXAMPLE 2. The figure eight. Let A and B be two circles in $\mathbb{R}^2$ with a point e in common, as in the figure. Let X be their union $X = A \vee B$. Let $f: X \to X$ be a map such that $f_\pi(\alpha) = \alpha^{-1}$ and $f_\pi(\beta) = \beta^2$, where α and β are the elements in the fundamental group shown in the figure, and $\mathrm{Fix}(f) = \{e,a\}$. Thus $L(f) = 0$, because the matrix of f_{1*} is $\begin{pmatrix} -1 & 0 \\ 0 & 2 \end{pmatrix}$. But it is obvious that $\operatorname{index}(f,a) = 1$, hence $\operatorname{index}(f,e) = -1$.

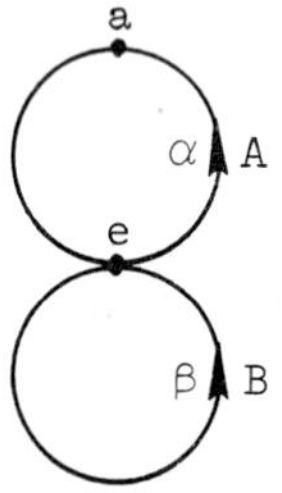

The two fixed points e and a are not in the same class. Hence $N(f) = 2$. Proof: Let w be the path from e to a on the right semicircle of A. Then $\langle w(f \circ w)^{-1} \rangle = \alpha$. Suppose there were a path u from e to a such that $u \simeq f \circ u$. Let $\gamma \in \pi_1(X,e)$ be the loop class of uw^{-1}. Then $\gamma \alpha f_\pi(\gamma^{-1}) = \langle uw^{-1} \rangle \alpha \langle (f \circ w)(f \circ u^{-1}) \rangle = \langle u(f \circ u^{-1}) \rangle = 1$. In the homology group $H_1(X)$, let $\gamma = k\alpha + \ell\beta$. Then $\begin{pmatrix} k \\ \ell \end{pmatrix} + \begin{pmatrix} 1 \\ 0 \end{pmatrix} - \begin{pmatrix} -1 & 0 \\ 0 & 2 \end{pmatrix}\begin{pmatrix} k \\ \ell \end{pmatrix} = \begin{pmatrix} 0 \\ 0 \end{pmatrix}$, i.e. $\begin{pmatrix} 2k+1 \\ -\ell \end{pmatrix} = \begin{pmatrix} 0 \\ 0 \end{pmatrix}$, which is absurd since $2k + 1 \neq 0 \mod 2$.

4.5 THEOREM. (Homotopy invariance). Let X be a connected, compact polyhedron, $H = \{h_t\}: f_0 \simeq f_1 : X \to X$ be a homotopy, and $\mathbb{F}_i$ be a fixed point class of f_i, $i = 0,1$. If $\mathbb{F}_0$ corresponds to $\mathbb{F}_1$ via H, then

$$\operatorname{index}(f_0, \mathbb{F}_0) = \operatorname{index}(f_1, \mathbb{F}_1) .$$

PROOF. Consider the fat homotopy given by H, that is, $\mathbb{H}: X \times I \to X \times I$. According to Theorem 2.7, $\mathbb{F}_0$ and $\mathbb{F}_1$ are the 0- and 1-slice of a fixed point class $\mathbb{F}$ of $\mathbb{H}$. Hence, by Corollary 3.10, we have

$\mathrm{index}(f_s, \mathbb{F}_s) = \mathrm{index}(h_s, \mathbb{F}_s) = \mathrm{index}(\mathbb{H}, \mathbb{F}), \quad s = 0,1.$ □

4.6 THEOREM. $N(f)$ is a homotopy invariant of f. □

4.7 THEOREM. Every map homotopic to f has at least $N(f)$ fixed points. in other words, $N(f) \leq \mathrm{Min}\{\#\mathrm{Fix}(g) \mid g \simeq f\}$. □

It is this theorem that made the Nielsen number so important in fixed point theory. More information on the significance of $N(f)$ will be presented in the next two sections.

Theorem 4.5 tells us that an essential fixed point class can never disappear (i.e. become empty) via a homotopy. We have an even stronger connectivity property. (Compare 2.8.)

4.8 THEOREM. Let $H : f_0 \simeq f_1 : X \to X$ be a homotopy and $\mathbb{F}_0$ be an essential fixed point class of f_0. Then, on $X \times I$ there exists a continuum C of fixed points of the fat homotopy $\mathbb{H}$ which intersects $\mathbb{F}_0 \times 0$ and $\mathbb{F}_1 \times 1$, where $\mathbb{F}_1$ is the fixed point class of f_1 corresponding to $\mathbb{F}_0$ via H.

PROOF. Let $\mathbb{F}$ be the fixed point class of $\mathbb{H}$, of which $\mathbb{F}_0$ and $\mathbb{F}_1$ are the 0- and 1-slice respectively. We need an elementary fact:

LEMMA. Let K be a compact metric space and let A and B be two disjoint closed subsets. Then, either (1) some component of K intersects both A, and B, or (2) K splits into a disjoint union of two closed subsets A' and B' with $A \subset A'$ and $B \subset B'$. (See [Whyburn], p. 12.)

Applying this lemma to the compact space $\mathbb{F}$ and its subsets $\mathbb{F}_0 \times 0$ and $\mathbb{F}_1 \times 1$, we only have to rule out the second alternative, namely the splitting of $\mathbb{F}$ into the disjoint union of two closed subsets S and T with $\mathbb{F}_0 \times 0 \subset S$ and $\mathbb{F}_1 \times 1 \subset T$. Suppose this were the case. Then S is an isolated set of fixed points of $\mathbb{H}$, hence by 3.10 we have

$$\mathrm{index}(h_0, S_0) = \mathrm{index}(\mathbb{H}, S) = \mathrm{index}(h_1, S_1) .$$

But $\mathrm{index}(h_0, S_0) = \mathrm{index}(f_0, \mathbb{F}_0) \neq 0$, while $\mathrm{index}(h_1, S_1) = 0$ because S_1 is empty. A contradiction. □

5. THE COMMUTATIVITY. Various commutativity properties in fixed point theory (e.g. the commutativity of the fixed point index) are extremely powerful. They permit us to switch from one space to another. "Commutativity" is a built-in symmetry of the fixed point problem. If we have $X \xrightarrow{f} Y$ and $Y \xrightarrow{g} X$, to find a fixed point of $g \circ f$ is to solve an equation $x = g \circ f(x)$, which is equivalent to a system of equations

$$\begin{cases} x = g(y) \\ y = f(x) . \end{cases}$$

This system is symmetric in x and y. Geometrically speaking, finding a fixed point of $g \circ f$ is equivalent to finding an intersection of the two graphs $\Gamma(f)$ and $\Gamma(g)$ in $X \times Y$.

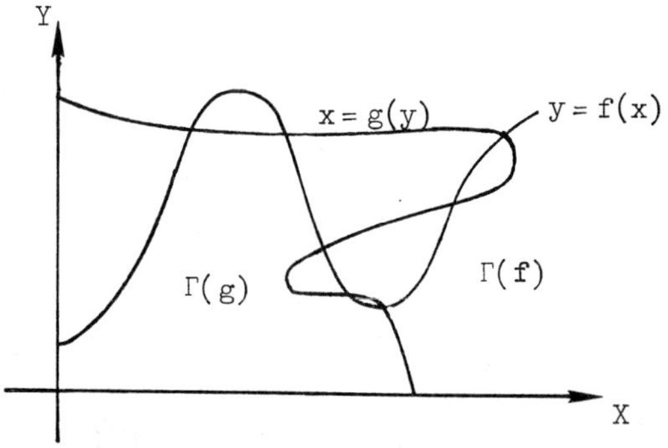

This is again a symmetric situation.

Now let us study this symmetry for fixed point classes. Let X, Y be compact connected polyhedra, $X \underset{g}{\overset{f}{\rightleftarrows}} Y$ be maps.

5.1 LEMMA. The maps $X \underset{g}{\overset{f}{\rightleftarrows}} Y$ restrict to a pair of homeomorphisms on the fixed point sets

$$\mathrm{Fix}(g \circ f) \underset{g}{\overset{f}{\rightleftarrows}} \mathrm{Fix}(f \circ g)$$

and they respect fixed point classes.

PROOF. The first part is obvious. We will show that if x_0 and x_1 are in the same fixed point class of $g \circ f$, then $f(x_0)$ and $f(x_1)$ are in the same fixed point class of $f \circ g$. In fact, if a path c from x_0 to x_1 is such that $c \simeq g \circ f \circ c$, then for the path $f \circ c$ from $f(x_0)$ to $f(x_1)$ we have $f \circ c \simeq f \circ (g \circ f \circ c) = (f \circ g) \circ (f \circ c)$, which ties $f(x_0)$ and $f(x_1)$ together. □

5.2 THEOREM (Commutativity of the Nielsen number and the indices of fixed point classes). Let $\mathbb{F}$ be a (non-empty) fixed point class of $g \circ f$, then $f(\mathbb{F})$ is a fixed point class of $f \circ g$, and

$$\mathrm{index}(g \circ f, \mathbb{F}) = \mathrm{index}(f \circ g, f(\mathbb{F})) .$$

Thus $N(g \circ f) = N(f \circ g)$.

PROOF. This is an easy consequence of Corollary 3.9. □

Theorem 5.2 enables us to show a stronger homotopy invariance of the Nielsen number, namely homotopy type invariance.

5.3 DEFINITION. Two maps $X \xrightarrow{f} X$ and $Y \xrightarrow{g} Y$ are said to be of the same <u>homotopy type</u> if there is a homotopy equivalence $h : X \to Y$ such that $h \circ f \simeq g \circ h$. Thus, for k a homotopy inverse of h, the following diagram is homotopy commutative:

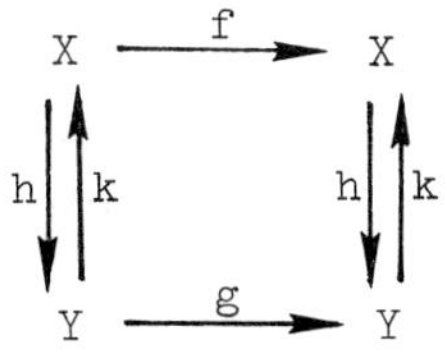

5.4 THEOREM (Homotopy type invariance of the Nielsen number). If X, Y are compact connected polyhedra, and $X \xrightarrow{f} X$ and $Y \xrightarrow{g} Y$ are of the same homotopy type, then $N(f) = N(g)$.

PROOF. Let $X \underset{k}{\overset{h}{\rightleftarrows}} Y$ be a homotopy equivalence as required in the definition above. Then by Theorem 5.2

$$N(f) = N((k \circ h) \circ f) = N(k \circ (h \circ f)) = N((h \circ f) \circ k)$$

$$= N((g \circ h) \circ k) = N(g \circ (h \circ k)) = N(g) . \qquad \square$$

Another useful consequence of Theorem 5.2 is

5.5 COROLLARY. Let X, X_0 be a pair of compact connected polyhedra, $X_0 \subset X$. Let $f : X \to X$ satisfy $f(X) \subset X_0$, and let $f_0 : X_0 \to X_0$ be its restriction. Then $N(f) = N(f_0)$.

PROOF. Apply Theorem 5.2 to $X \underset{i}{\overset{f}{\rightleftarrows}} X_0$, where i is the inclusion. $\square$

Note that X and X_0 may have different fundamental groups. For example if $X_0 \simeq S^1$, then this will reduce the computation of $N(f)$ on a complicated space X to computation on a nice space S^1.

EXAMPLE 1. If $X \xrightarrow{f} X$ is such that $f^2 = f$, then f is a retraction of X onto $f(X)$. Since $\mathrm{Fix}(f) = f(X)$ is connected, there is only one non-empty fixed point class, hence $N(f) \leq 1$.

EXAMPLE 2. Let $X \underset{d}{\overset{u}{\leftrightarrows}} Y$ be a domination by a compact polyhedron X of a compact polyhedron Y, that is, $d \circ u \simeq \mathrm{id}_Y$. Let $f = u \circ d : X \to X$. Then $f^2 \simeq f$ and $N(f) = N(u \circ d) = N(d \circ u) = N(\mathrm{id}_Y) \leq 1$.

5.6 CONJECTURE (Geoghegan, 1979). If $f : X \to X$ is a homotopy idempotent (in the sense that $f \simeq f^2$), then $N(f) \leq 1$.

This conjecture is related to the converse of the last example: does every homotopy idempotent on a finite complex come from a domination over another finite complex? For more details, see the paper [Geoghegan].

5.7 REMARK. In 5.2 we have only considered the relation between <u>nonempty</u> fixed point classes of $g \circ f$ and $f \circ g$. It is true that there is a 1-1 correspondence between lifting classes of $g \circ f$ and those of $f \circ g$. This will be proved later in §III.1, to avoid unnecessary repetition of similar arguments. Here we only remark that it is impossible to establish a 1-1 correspondence on the lifting level, since $\tilde{X}$ and $\tilde{Y}$ may have different numbers of sheets.

6. THE LEAST NUMBER OF FIXED POINTS. Let X be a compact connected polyhedron, and let $f: X \to X$ be a map. Consider the number

$$MF[f] := \operatorname{Min}\{\#\operatorname{Fix}(g) \mid g \simeq f\},$$

i.e. the least number of fixed points in the homotopy class $[f]$ of f. We know from §4 that $N(f)$ is a lower bound for $MF[f]$. The importance of the Nielsen number in fixed point theory lies in the fact that, under a mild restriction on the space involved, it is indeed the minimal number of fixed points in the homotopy class.

The number $MF[f]$ is geometrically defined. There is no hint in the definition of how to compute it. So, the relationship between $MF[f]$ and $N(f)$ is significant. The equality $N(f) = MF[f]$ means we can homotope the map f so that each essential fixed point class is combined into a single fixed point and each inessential fixed point class is removed. This problem is geometric in nature and can be solved only by careful constructions, for which we need local restrictions on the space.

To pursue this geometric problem we would need a separate chapter. Instead, we will only state some important results and unsolved questions, and recommend the paper [Jiang (1980)] to the interested reader. In the later chapters we will concentrate on the more algebraic part of the theory of fixed point classes, especially the problem of computation.

6.1 DEFINITION. A point x of a connected space X is a (global) separating point of X if $X - x$ is not connected. A point x of a space X is a local separating point if x is a separating point of some connected open subspace U of X.

6.2 THEOREM. Let X be a compact connected polyhedron. Suppose X is not disconnected by removing all local separating points. Then $MF[id_X] = N(id_X)$, which is 0 if $\chi(X) = 0$ and is 1 if $\chi(X) \neq 0$.

6.3 THEOREM. Let X be a compact connected polyhedron without local separating points. Suppose X is not a surface (closed or with boundary) of negative Euler characteristic. Then $MF[f] = N(f)$ for any map $f: X \to X$.

QUESTION 1. Let X be a compact surface with $\chi(X) < 0$. Is it still true that $MF[f] = N(f)$ for every $f: X \to X$?

QUESTION 2. Let M be a closed manifold, and let $\varphi: X \to X$ be a homeomorphism. Is it true that

$$N(\varphi) = \mathrm{Min}\{\#\mathrm{Fix}(\psi) \mid \psi \text{ is isotopic to } \varphi\} \ ?$$

It is known to be true when $\dim M \leq 2$.

We now give two consequences of Theorem 6.3.

6.4 THEOREM (Geometric characterization of the Nielsen number). In the category of compact connected polyhedra, the Nielsen number of a self-map equals the least number of fixed points among all self-maps having the same homotopy type.

PROOF. Let $f: X \to X$ be a self-map in the category, and consider the number

$$m = \mathrm{Min}\{\#\mathrm{Fix}(g) \mid g \text{ has the same homotopy type as } f\} .$$

Then $N(f) \leq m$ by Theorems 5.4 and 4.3. On the other hand, there always exists a manifold M (with boundary) of dimension ≥ 3 which has the same homotopy type as X (for example, M = the regular neighborhood of X imbedded in a Euclidean space). So, by Theorem 6.3, the lower bound $N(f)$ is realizable on M by a map having the same homotopy type as f. □

6.5 THEOREM. In the category of compact connected polyhedra without global separating points, the fixed point property (i.e. the property that every self-map has a fixed point) is a homotopy type invariant.

PROOF. Let us say a space X has Property (N) if $N(f) > 0$ for every $f: X \to X$. This property is a homotopy type invariant, since $N(f)$ is a homotopy type invariant by Theorem 5.4. Clearly Property (N) implies the fixed point property. We claim that in the category specified, the fixed point property implies Property (N) as well.

If X has no local separating points and is not a surface of negative Euler characteristic, this follows from Theorem 6.3. If X has no global separating point but does have a local one (cf. Lemma II.6.4), or if X is a surface with $\chi < 0$, then X has S^1 as a retract, hence can never have the fixed point property. This proves our claim. □

CHAPTER II

COMPUTATION OF THE NIELSEN NUMBER

Although the Nielsen number plays an important role theoretically, its computation is no easy task. We will give an exposition on the progress in this direction. Coordinates are introduced in §1 as the basis for algebraization. It turns out that the homotopy invariance theorem I.4.5 is quite helpful in computation. Cyclic homotopies are discussed in §3 and applied in §4 to a fairly large class of spaces. For compact polyhedra with a finite fundamental group, the Lefschetz fixed point theorem is applicable on the universal coverings. This case is studied in §5. Finally, the results of this chapter are combined with those in §I.6 to give converses of the Lefschetz fixed point theorem.

1. COORDINATES FOR LIFTING CLASSES. Let $f : X \to X$ be given, and a specific lifting $\tilde{f} : \tilde{X} \to \tilde{X}$ be chosen as reference. Then every lifting of f can be uniquely written as $\alpha \circ \tilde{f}$, with $\alpha \in \mathcal{D}$, according to Proposition I.1.2. So elements of $\mathcal{D}$ serve as coordinates of liftings with respect to the reference $\tilde{f}$.

Now for every $\alpha \in \mathcal{D}$, the composition $\tilde{f} \circ \alpha$ is also a lifting of f, so there is a unique element $\alpha' \in \mathcal{D}$ such that $\alpha' \circ \tilde{f} = \tilde{f} \circ \alpha$. This correspondence $\alpha \to \alpha'$ is determined by the reference $\tilde{f}$, and is obviously a homomorphism.

1.1 DEFINITION. The endomorphism $\tilde{f}_\pi : \mathcal{D} \to \mathcal{D}$ determined by a lifting $\tilde{f}$ of f is defined by

$$\tilde{f}_\pi(\alpha) \circ \tilde{f} = \tilde{f} \circ \alpha, \qquad \alpha \in \mathcal{D}$$

It is well known that $\mathcal{D} \cong \pi_1(X)$. We will identify these two in the following way.

1.2 IDENTIFICATION. Pick base points $x_0 \in X$ and $\tilde{x}_0 \in p^{-1}(x_0) \subset \tilde{X}$ once for all. Now points of $\tilde{X}$ are in 1-1 coorrespondence with path classes in X starting from x_0: For $\tilde{x} \in \tilde{X}$, take any path in $\tilde{X}$ from $\tilde{x}_0$ to $\tilde{x}$ and project it into X; for a path c in X starting from x_0, lift it into $\tilde{X}$ from $\tilde{x}_0$ and get its endpoint. In this way, we identify a point of $\tilde{X}$ with a path class $\langle c \rangle$ in X starting from x_0. Note that under this identification, $\tilde{x}_0 = \langle e \rangle$, the unit element in $\pi_1(X,x_0)$. The action of the loop class $\alpha = \langle a \rangle \in \pi_1(X,x_0)$ on $\tilde{X}$ is then given by

$$\alpha = \langle a \rangle : \langle c \rangle \mapsto \alpha\langle c \rangle = \langle ac \rangle \ .$$

Every lifting $\tilde{f}$ is (according to Proposition I.1.2 again) uniquely determined by its value, $\tilde{f}(\tilde{x}_0) \in p^{-1}(f(x_0))$. Suppose $\tilde{f}(\tilde{x}_0) = \langle w \rangle$, where w is a path from x_0 to $f(x_0)$. Then for any point $\langle c \rangle \in \tilde{X}$, we have

$$\tilde{f} : \langle c \rangle \mapsto \langle w(f \circ c) \rangle$$

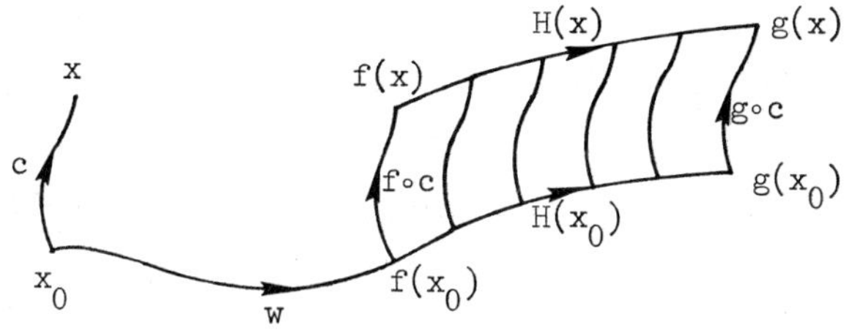

A homotopy $H : f \simeq g$ lifts to a homotopy $\tilde{H} : \tilde{f} \simeq \tilde{g}$, then $\tilde{g}$ is determined by its value $\tilde{g}(\tilde{x}_0) = \langle wH(x_0) \rangle$, where $H(x_0)$ is the <u>trace</u> $\{h_t(x_0)\}$ of H. Thus,

$$\tilde{g} : \langle c \rangle \mapsto \langle wH(x_0)(f \circ c) \rangle \ .$$

Now that we have identified $\mathcal{D}$ with $\pi = \pi_1(X,x_0)$, we should clarify the relationship between $\pi \xrightarrow{\tilde{f}_\pi} \pi$ and $\pi_1(X,x_0) \xrightarrow{f_\pi} \pi_1(X,f(x_0))$.

1.3 LEMMA. Suppose $\tilde{f}(\tilde{x}_0) = \langle w \rangle$. Then the following diagram commutes:

$$\begin{array}{ccccc} \pi_1(X,x_0) & \xrightarrow{f_\pi} & \pi_1(X,f(x_0)) & \xrightarrow{w_*} & \pi_1(X,x_0) \\ \big\downarrow & & & & \big\uparrow \\ \longrightarrow & & \tilde{f}_\pi & & \longrightarrow \end{array}$$

where f_π is the homomorphism induced by the map f and w_* is the isomorphism induced by the path w. In other words, for every $\alpha = \langle a \rangle \in \pi_1(X,x_0)$, we have

$$\tilde{f}_\pi(\langle a \rangle) = \langle w(f \circ a)w^{-1} \rangle \ .$$

In particular, if $x_0 \in p\,\mathrm{Fix}(\tilde{f})$ and $\tilde{x}_0 \in \mathrm{Fix}(\tilde{f})$, then $\tilde{f}_\pi = f_\pi$.

PROOF. Let $\alpha' = \langle a' \rangle = \tilde{f}_\pi(\alpha)$. Then by definition, $\alpha' \circ \tilde{f} = \tilde{f} \circ \alpha$. Since

$$\langle c\rangle \overset{\tilde f}{\longmapsto} \langle w(f\circ c)\rangle \overset{\alpha'}{\longmapsto} \langle a'\rangle\langle w(f\circ c)\rangle ,$$

$$\langle c\rangle \overset{\alpha}{\longmapsto} \langle ac\rangle \overset{\tilde f}{\longmapsto} \langle w(f\circ (ac))\rangle = \langle w(f\circ a)(f\circ c)\rangle$$

$$= \langle w(f\circ a)w^{-1}\rangle\langle w(f\circ c)\rangle ,$$

we have $\langle a'\rangle = \langle w(f\circ a)w^{-1}\rangle$, that is $\tilde f_\pi(\alpha) = w_* f_\pi(\alpha)$. □

We have seen that $\alpha \in \pi$ can be considered as the coordinate of the lifting $\alpha\circ\tilde f$. Can we tell the conjugacy of two liftings from their coordinates?

1.4 LEMMA. $[\alpha\circ\tilde f] = [\alpha'\circ\tilde f]$ iff there is $\gamma\in\pi$ such that $\alpha' = \gamma\alpha\tilde f_\pi(\gamma^{-1})$.

PROOF. $[\alpha\circ\tilde f] = [\alpha'\circ\tilde f]$ iff there is $\gamma\in\pi$ such that $\alpha'\circ\tilde f = \gamma\circ(\alpha\circ\tilde f)\circ\gamma^{-1} = \gamma\alpha\tilde f_\pi(\gamma^{-1})\circ\tilde f$. □

1.5 DEFINITION. Let G be a group, $\varphi : G\to G$ an endomorphism. Two elements $\alpha, \alpha'\in G$ are said to be φ-conjugate iff there exists $\gamma\in G$ such that $\alpha' = \gamma\alpha\varphi(\gamma^{-1})$.

We will write $[\alpha]$ for the φ-conjugacy class of $\alpha\in G$ in case the homomorphism φ in question is clear from the context.

The following are simple but very useful facts about φ-conjugacy classes.

1.6 LEMMA. For any $\alpha, \beta\in G$, we have

(i) $\alpha\beta \sim \beta\varphi(\alpha)$.

(ii) $\alpha\sim\varphi(\alpha)$,

where "$\sim$" denotes the φ-conjugacy relation.

PROOF. (i) $\alpha\beta\sim\alpha^{-1}(\alpha\beta)\varphi(\alpha) = \beta\varphi(\alpha)$.

(ii) Take $\beta = 1$ in (i). □

1.7 THEOREM. Lifting classes of f are in 1-1 correspondence with $\tilde f_\pi$-conjugacy classes in π, the lifting class $[\gamma\circ\tilde f]$ corresponding to the $\tilde f_\pi$-conjugacy class of γ. □

By an abuse of language, we will say that the fixed point class $p\,\mathrm{Fix}(\gamma\circ\tilde f)$, which is labeled with the lifting class $[\gamma\circ\tilde f]$, corresponds to the $\tilde f_\pi$-conjugacy class of γ. Thus the $\tilde f_\pi$-conjugacy classes in π serve as coordinates for the fixed point classes of f, once a reference lifting $\tilde f$ is chosen. We state this as a definition.

1.8 DEFINITION. Let a reference lifting $\tilde f$ of f be chosen. Then the $\tilde f_\pi$-conjugacy class of $\gamma\in\pi$ is said to be the coordinate for the lifting class $[\gamma\circ\tilde f]$ and the fixed point class $p\,\mathrm{Fix}(\gamma\circ\tilde f)$.

When the base point is chosen to be in the fixed point class labeled by the reference lifting $\tilde f$, the coordinate of a fixed point class can be obtained geometrically.

1.9 COROLLARY. Suppose the base points are such that $x_0 \in p\,\mathrm{Fix}(\tilde{f})$ and $\tilde{x}_0 \in \mathrm{Fix}(\tilde{f})$, where $\tilde{f}$ is the reference lifting. Then the coordinate for the class of a fixed point x of f is the $\tilde{f}_\pi$-conjugacy class of $\gamma = \langle c(f \circ c)^{-1}\rangle \in \pi$, where c is any path from x_0 to x. In other words, $x \in p\,\mathrm{Fix}(\gamma \circ \tilde{f})$.

PROOF. Let $\tilde{x} = \langle c\rangle \in p^{-1}(x)$. We only have to show $\tilde{x} \in \mathrm{Fix}(\gamma \circ \tilde{f})$. Since $\tilde{f}(\tilde{x}_0) = \tilde{x}_0 = \langle e\rangle$, we have $\tilde{f}(\tilde{x}) = \langle e(f \circ c)\rangle = \langle f \circ c\rangle$. Hence $(\gamma \circ \tilde{f})(\tilde{x}) = \gamma\langle f \circ c\rangle = \langle c(f \circ c)^{-1}(f \circ c)\rangle = \langle c\rangle = \tilde{x}$, or $\tilde{x} \in \mathrm{Fix}(\gamma \circ \tilde{f})$. □

2. A LOWER BOUND FOR THE REIDEMEISTER NUMBER. A reasonable approach is to consider homomorphisms of π which send an $\tilde{f}_\pi$-conjugacy class into an element.

2.1 THEOREM. Let $\tilde{f}$ be a lifting of $f : X \to X$. Then the composition $\eta \circ \theta$

$$\pi = \pi_1(X,x_0) \xrightarrow{\theta} H_1(X) \xrightarrow{\eta} \mathrm{Coker}(H_1(X) \xrightarrow{1-f_{1*}} H_1(X)) ,$$

where θ is abelianization and η is the natural projection, sends every $\tilde{f}_\pi$-conjugacy class into a single element. Moreover, any homomorphism $\zeta : \pi \to G$, from π to a group G, with the property that it sends every $\tilde{f}_\pi$-conjugacy class into a single element, factors through $\eta \circ \theta$.

PROOF. From Lemma 1.3 we have a commutative square

$$\begin{array}{ccc} \pi & \xrightarrow{\tilde{f}_\pi} & \pi \\ \theta\big\downarrow & & \big\downarrow\theta \\ H_1(X) & \xrightarrow{f_{1*}} & H_1(X) \end{array}$$

If $\alpha' = \gamma\alpha\tilde{f}_\pi(\gamma^{-1})$, then

$$\theta(\alpha') = \theta(\gamma) + \theta(\alpha) + \theta(\tilde{f}_\pi(\gamma^{-1})) = \theta(\gamma) + \theta(\alpha) - f_{1*}(\theta(\gamma))$$

$$= \theta(\alpha) + (1 - f_{1*})\theta(\gamma) ,$$

hence $\eta \circ \theta(\alpha) = \eta \circ \theta(\alpha')$. This proves the first part of the theorem.

Now let $\zeta : \pi \to G$ be a homomorphism such that ζ sends every $\tilde{f}_\pi$-conjugacy class into a single element of G. By Lemma 1.6 (ii), we have $\zeta(\alpha) = \zeta(\tilde{f}_\pi(\alpha))$, and then by 1.6 (i) we have

$$\zeta(\alpha\beta) = \zeta(\beta)\zeta(\tilde{f}_\pi(\alpha)) = \zeta(\beta)\zeta(\alpha) = \zeta(\beta\alpha) ,$$

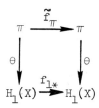

for any $\alpha, \beta \in \pi$. Thus ζ factors through the abelianization θ, i.e. there is a homomorphism $\bar{\zeta} : H_1(X) \to G$ such that $\zeta = \bar{\zeta} \circ \theta$. Every element in $\mathrm{Im}(1 - f_{1*})$ can be written as $(1 - f_{1*})\theta(\alpha) = \theta(\alpha) - f_{1*} \circ \theta(\alpha)$ $= \theta(\alpha) - \theta \circ \tilde{f}_\pi(\alpha) = \theta(\alpha\tilde{f}_\pi(\alpha^{-1}))$, hence $\bar{\zeta}((1 - f_{1*})\theta(\alpha)) = \bar{\zeta} \circ \theta(\alpha\tilde{f}_\pi(\alpha^{-1}))$ $= \zeta(\alpha\tilde{f}_\pi(\alpha^{-1})) = \zeta(\alpha)\zeta(\tilde{f}_\pi(\alpha))^{-1} = \zeta(\alpha)\zeta(\alpha)^{-1} = 1$, in other words $\mathrm{Im}(1 - f_{1*}) \subset \mathrm{Ker}\,\bar{\zeta}$. Thus $\bar{\zeta}$ factors through η, therefore $\zeta = \bar{\zeta} \circ \theta$ factors through $\eta \circ \theta$. □

This theorem shows the importance of the group $\mathrm{Coker}(1 - f_{1*})$. For example,

2.2 COROLLARY. $R(f) \geq \#\mathrm{Coker}(1 - f_{1*})$. □

EXERCISE. $R(\mathrm{id}_X) \geq$ the order of the center of π.

2.3 THEOREM. The following conditions are equivalent, where "$\sim$" denotes the $\tilde{f}_\pi$-conjugacy relation in π.

(i) The epimorphism $\eta \circ \theta : \pi \to \mathrm{Coker}(1 - f_{1*})$ maps different $\tilde{f}_\pi$-conjugacy classes into different elements. (Hence $R(f) = \#\mathrm{Coker}(1 - f_{1*})$.)

(ii) The relation $\alpha \sim \alpha'$ implies $\alpha\beta \sim \alpha'\beta$ for any $\beta \in \pi$. (It also implies $\beta\alpha \sim \beta\alpha'$ for any $\beta \in \pi$, and $\alpha^{-1} \sim \alpha'^{-1}$.)

(iii) For any $\alpha, \beta, \gamma \in \pi$, the relation $\alpha\beta\gamma \sim \beta\alpha\gamma$ holds.

PROOF. (i) $\Rightarrow$ (ii):

$$\begin{aligned} \alpha \sim \alpha' &\Rightarrow \eta \circ \theta(\alpha) = \eta \circ \theta(\alpha') && \text{by Theorem 2.1} \\ &\Rightarrow \eta \circ \theta(\alpha\beta) = \eta \circ \theta(\alpha'\beta) && \text{since } \eta \circ \theta \text{ is a homomorphism} \\ &\Rightarrow \alpha\beta \sim \alpha'\beta && \text{by (i).} \end{aligned}$$

(The other two implications are similar.)

(ii) $\Rightarrow$ (iii):

$$\begin{aligned} \alpha\beta \sim \tilde{f}_\pi(\alpha\beta) = \tilde{f}_\pi(\alpha)\tilde{f}_\pi(\beta) \text{ and } \tilde{f}_\pi(\alpha) \sim \alpha && \text{by 1.6} \\ \Rightarrow \alpha\beta \sim \tilde{f}_\pi(\alpha)\tilde{f}_\pi(\beta) \sim \alpha\tilde{f}_\pi(\beta) && \text{by (ii)} \\ \sim \beta\alpha && \text{by 1.6} \\ \Rightarrow \alpha\beta\gamma \sim \beta\alpha\gamma && \text{by (ii).} \end{aligned}$$

(iii) $\Rightarrow$ (i): We want to show that if $\eta \circ \theta(\alpha) = \eta \circ \theta(\alpha')$ then $\alpha \sim \alpha'$. There are three steps.

(a) For any commutator $[\alpha,\beta] = \alpha\beta\alpha^{-1}\beta^{-1}$ and any γ,

$$[\alpha,\beta]\gamma = \alpha\beta(\alpha^{-1}\beta^{-1}\gamma) \sim \beta\alpha(\alpha^{-1}\beta^{-1}\gamma) = \gamma \qquad \text{by (iii).}$$

(b) If $\theta(\gamma) = \theta(\gamma')$, then $\gamma \sim \gamma'$.
In fact, $\theta(\gamma'\gamma^{-1}) = 0$ implies $\gamma'\gamma^{-1}$ is a product of commutators, i.e.

$$\gamma' = [\alpha_1,\beta_1]\cdots[\alpha_k,\beta_k]\gamma .$$

Applying (a) repeatedly, we get $\gamma' \sim \gamma$.

(c) Now assume $\eta \circ \theta(\alpha) = \eta \circ \theta(\alpha')$. By the definition of η, there exists $c \in H_1(X)$ and $\gamma \in \theta^{-1}(c)$, such that

$$\theta(\alpha') - \theta(\alpha) = (1 - f_{1*})(c) = \theta(\gamma) - f_{1*}\theta(\gamma) = \theta(\gamma) - \theta(\tilde{f}_\pi(\gamma))$$

$$\Longleftrightarrow \theta(\alpha') = \theta(\gamma) + \theta(\alpha) - \theta(f_\pi(\gamma)) = \theta(\gamma\alpha\tilde{f}_\pi(\gamma^{-1}))$$

$$\Rightarrow \alpha' \sim \gamma\alpha\tilde{f}_\pi(\gamma^{-1}) \sim \alpha \qquad \text{(by (b))}. \qquad \square$$

2.4 DEFINITION. A map $f : X \to X$ is said to be <u>eventually commutative</u> if there exists a natural number n such that $(f^n)_\pi\pi_1(X,x_0)$ $(\subset \pi_1(X,f^n(x_0)))$ is commutative.

By means of Lemma 1.3, it is easily seen that $(f^n)_\pi\pi$ is abelian iff $(\tilde{f}_\pi)^n\pi$ is abelian, for any reference lifting $\tilde{f}$ of f.

2.5 THEOREM. If f is eventually commutative, then $R(f) = \#\mathrm{Coker}(1 - f_{1*})$.

PROOF. It suffices to verify condition (iii) of Theorem 2.3. In fact,

$$\alpha\beta\gamma \sim \tilde{f}^n_\pi(\alpha\beta\gamma) = \tilde{f}^n_\pi(\beta\alpha\gamma) \sim \beta\alpha\gamma . \qquad \square$$

EXERCISE. Determine $\#\mathrm{Coker}(1 - f_{1*})$ in the following cases:

1. $H_1(X)$ is free abelian of rank n and f_{1*} is given by an $n \times n$ integral matrix A.
2. $H_1(X)$ is cyclic of order m and f_{1*} is multiplication by d.

3. THE TRACE SUBGROUP OF CYCLIC HOMOTOPIES. The homotopy invariance Theorem I.4.5. tells us that if a homotopy $\{h_t\} : f \simeq g : X \to X$ lifts to $\{\tilde{h}_t\} : \tilde{f} \simeq \tilde{g} : \tilde{X} \to \tilde{X}$, then

$$\mathrm{index}(f,p\,\mathrm{Fix}(\tilde{f})) = \mathrm{index}(g,p\,\mathrm{Fix}(\tilde{g})) .$$

Suppose $\{h_t\}$ is a cyclic homotopy $\{h_t\} : f \simeq f$, then it lifts to a homotopy from a given lifting $\tilde{f}$ to another lifting $\tilde{f}' = \alpha \circ \tilde{f}$, and we have

$$\mathrm{index}(f,p\,\mathrm{Fix}(\tilde{f})) = \mathrm{index}(f,p\,\mathrm{Fix}(\alpha \circ \tilde{f})) .$$

In other words, a cyclic homotopy induces a permutation among the lifting classes (hence the fixed point classes); those on the same orbit of this permutation have the same index. We discuss this permutation group in this section, then apply this idea to the computation of $N(f)$ in the next section.

3.1 DEFINITION. Let $f : X \to X$ be a self-map of a compact connected polyhedron, and let $\tilde{f} : \tilde{X} \to \tilde{X}$ be a lifting of f. Define

$$J(\tilde{f}) := \{\alpha \in \pi \mid \text{there exists a cyclic homotopy } \{h_t\} : f \simeq f \text{ which lifts to } \{\tilde{h}_t\} : \tilde{f} \simeq \alpha \circ \tilde{f}\} .$$

3.2 PROPOSITION. $J(\tilde{f}) \subset \pi$ is a subgroup.

PROOF. Suppose $\alpha, \beta \in J(\tilde{f})$, that is, there exist $\{h_t\} : f \simeq f$ and $\{h'_t\} : f \simeq f$ which lift to $\{\tilde{h}_t\} : \tilde{f} \simeq \alpha \circ \tilde{f}$ and $\{\tilde{h}'_t\} : \tilde{f} \simeq \beta \circ \tilde{f}$. Then $\{h''_t\} = \{h'_{1-t}\}$ lifts to $\{\tilde{h}''_t\} = \{\alpha\beta^{-1} \circ \tilde{h}'_{1-t}\} : \alpha \circ \tilde{f} = (\alpha\beta^{-1}) \circ (\beta \circ \tilde{f}) \simeq \alpha\beta^{-1} \circ \tilde{f}$. So the multiplication $\{h_t\} \cdot \{h''_t\} : f \simeq f$ lifts to $\{\tilde{h}_t\} \cdot \{\tilde{h}''_t\} : \tilde{f} \simeq \alpha\beta^{-1} \circ \tilde{f}$, hence $\alpha\beta^{-1} \in J(\tilde{f})$. □

3.3 LEMMA. $J(\tilde{f}) \subset Z(\tilde{f}_\pi(\pi),\pi)$. In particular, $J(\mathrm{id}_{\tilde{X}}) \subset Z(\pi)$.

Here $Z(G)$ denotes the center of a group G, and $Z(H,G)$ denotes the centralizer of a subgroup $H \subset G$, i.e. $Z(H,G) = \{g \in G \mid gh = hg \text{ for all } h \in H\}$.

PROOF. Let $\alpha \in J(\tilde{f})$ and $\xi \in \pi$. We want to show $\alpha\tilde{f}_\pi(\xi) = \tilde{f}_\pi(\xi)\alpha$. Suppose $\{h_t\} : f \simeq f$ lifts to $\{\tilde{h}_t\} : \tilde{f} \simeq \alpha \circ \tilde{f}$. Look at the lifting of $\{h_t\}$ starting from $\tilde{f} \circ \xi = \tilde{f}_\pi(\xi) \circ \tilde{f}$. On one hand,

$$\{\tilde{h}_t \circ \xi\} : \tilde{f} \circ \xi \simeq (\alpha \circ \tilde{f}) \circ \xi = \alpha \circ (\tilde{f} \circ \xi) = (\alpha \circ \tilde{f}_\pi(\xi)) \circ \tilde{f} = (\alpha\tilde{f}_\pi(\xi)) \circ \tilde{f} ,$$

but on the other hand,

$$\{\tilde{f}_\pi(\xi) \circ \tilde{h}_t\} : \tilde{f}_\pi(\xi) \circ \tilde{f} \simeq \tilde{f}_\pi(\xi) \circ (\alpha \circ \tilde{f}) = (\tilde{f}_\pi(\xi)\alpha) \circ \tilde{f} .$$

By the uniqueness of liftings, we have

$$\alpha\tilde{f}_\pi(\xi) = \tilde{f}_\pi(\xi)\alpha .$$ □

3.4 LEMMA. Suppose $f, g : X \to X$ are two maps, $\tilde{f}, \tilde{g} : \tilde{X} \to \tilde{X}$ are their liftings respectively. Then

$$J(\tilde{g}) \subset J(\tilde{g} \circ \tilde{f}) \quad \text{and} \quad \tilde{g}_\pi(J(\tilde{f})) \subset J(\tilde{g} \circ \tilde{f}) .$$

In particular, $J(\mathrm{id}_{\tilde{X}}) \subset J(\tilde{f})$.

PROOF. Suppose $\{h_t\} : f \simeq f$ lifts to $\{\tilde{h}_t\} : \tilde{f} \simeq \alpha \circ \tilde{f}$, and $\{\tilde{h}'_t\} : g \simeq g$ lifts to $\{h'_t\} : \tilde{g} \simeq \beta \circ \tilde{g}$. Then

$$\{\tilde{h}'_t \circ \tilde{f}\} : \tilde{g} \circ \tilde{f} \simeq (\beta \circ \tilde{g}) \circ \tilde{f} = \beta \circ (\tilde{g} \circ \tilde{f}) ,$$

$$\{\tilde{g} \circ \tilde{h}_t\} : \tilde{g} \circ \tilde{f} \simeq \tilde{g} \circ (\alpha \circ \tilde{f}) = (\tilde{g} \circ \tilde{\alpha}) \circ \tilde{f} = \tilde{g}_\pi(\alpha) \circ (\tilde{g} \circ \tilde{f}) .$$

Hence β and $g_\pi(\alpha)$ are in $J(\tilde{g} \circ \tilde{f})$. □

Now we make use of the identification 1.2 again. Suppose $\tilde{f}(\tilde{x}_0) = \langle w \rangle$. If a cyclic homotopy $H : f \simeq f$ lifts to a homotopy $\tilde{H} : \tilde{f} \simeq \alpha \circ \tilde{f}$, then $(\alpha \circ \tilde{f})(\tilde{x}_0) = \langle wH(x_0) \rangle = \langle wH(x_0)w^{-1} \rangle \langle w \rangle$, but $\alpha(\tilde{f}(\tilde{x}_0)) = \alpha(\langle w \rangle) = \alpha \langle w \rangle$, hence $\alpha = \langle wH(x_0)w^{-1} \rangle = w_*\langle H(x_0) \rangle$. This analysis leads us to the following notion.

3.5 DEFINITION. The trace subgroup of cyclic homotopies $J(f,x_0) \subset \pi_1(X,f(x_0))$ is defined by

$$J(f,x_0) := \{\xi \in \pi_1(X,f(x_0)) \mid \text{there exists a cyclic homotopy } H = \{h_t\} : f \simeq f \text{ such that } \langle H(x_0) \rangle = \xi\}$$

$$J(X) := J(\mathrm{id}_X,x_0) \subset \pi_1(X,x_0) = \pi .$$

That $J(f,x_0)$ is indeed a subgroup follows from 3.2 and the lemma below.

3.6 LEMMA. Suppose $\tilde{f}(\tilde{x}_0) = \langle w \rangle$, where w is a path from x_0 to $f(x_0)$. Then the isomorphism $w_* : \pi_1(X,f(x_0)) \to \pi_1(X,x_0) = \pi$ maps $J(f,x_0)$ onto $J(\tilde{f})$. In particular, $J(X) = J(\mathrm{id}_{\tilde{X}})$. □

Recall from Lemma 1.3 that $\tilde{f}_\pi = w_* \circ f_\pi$, hence $f_\pi^{-1}J(f,x_0) = \tilde{f}_\pi^{-1}J(\tilde{f})$. Lemmas 3.3 and 3.4 can be restated as

3.7 LEMMA. $J(f,x_0) \subset Z(f_\pi(\pi_1(X,x_0)),\pi_1(X,f(x_0)))$. In particular, $J(X) \subset Z(\pi)$. □

3.8 LEMMA. $J(g,f(x_0)) \subset J(g \circ f,x_0)$, $g_\pi(J(f,x_0)) \subset J(g \circ f,x_0)$. In particular, $J(\mathrm{id}_X,f(x_0)) \subset J(f,x_0)$. □

The advantage of $J(f,x_0)$ over $J(\tilde{f})$ is that it does not involve the covering space explicitly, hence is easier to handle. It is independent of the base point x_0 in the following sense.

3.9 LEMMA. Let w be a path in X from x_0 to x_1. Then the isomorphism $(f \circ w)_* : \pi_1(X,f(x_1)) \to \pi_1(X,f(x_0))$, induced by the path $f \circ w$, restricts to an isomorphism $(f \circ w)_* : J(f,x_1) \to J(f,x_0)$, and this latter isomorphism does not depend on the choice of w.

PROOF. Given a cyclic homotopy $H : f \simeq f$, it is obvious that $H(x_0) \simeq (f \circ w)(H(x_1))(f \circ w)^{-1}$, hence $\langle H(x_0) \rangle = (f \circ w)_*\langle H(x_1) \rangle$, for any path w from x_0 to x_1. □

3.10 COROLLARY. The conditions $J(\mathrm{id}_X,x_0) = \pi_1(X,x_0)$ and $f_\pi\pi_1(X,x_0) \subset J(f,x_0)$ are independent of the base point x_0. Hence we will write them as $J(X) = \pi_1(X)$ and $f_\pi\pi_1(X) \subset J(f)$ respectively. □

Lemma 3.8 tells us that $J(X)$ is contained in every $J(f)$. This leads us to consider spaces with $J(X) = \pi_1(X)$. From 3.7 we know that, for such spaces, $\pi_1(X)$ is abelian.

3.11 THEOREM. The class of path-connected spaces satisfying the condition $J(X) = \pi_1(X)$ is closed under homotopy equivalence and the topological product operation, and contains the following:

(i) simply connected spaces,

(ii) generalized lens spaces $L(m;q_1,\dots,q_n)$,

(iii) H-spaces,

(iv) homogeneous spaces of the form G/G_0 where G is a topological group, G_0 a subgroup which is a connected compact Lie group.

PROOF. This class of spaces is clearly closed under homotopy equivalence and is also closed under the topological product operation since $J(X \times Y) = J(X) \times J(Y)$.

Case (i) is trivial. Case (ii) is easy to see by a suitable rotation of S^{2n+1}.

Case (iii). Let (X,e) be an H-space with base point e, multiplication $\mu : X \times X \to X$, and a homotopy $F : \mathrm{id}_X \simeq \mu(e,\cdot) : X,e \to X,e$. (Cf. [Spanier] p. 35 for the definition.) For any element $\alpha \in \pi_1(X,e)$, pick a representative loop $\{x_t\}_{t\in I}$, consider the cyclic homotopy H consisting of F followed by $\{\mu(x_t,\cdot)\}_{t\in I} : X \to X$ followed by the reverse F^{-1} of F. Then evidently the trace $H(e)$ is in α. This shows $J(X,e) = \pi_1(X,e)$.

Case (iv). $X = G/G_0$ is the space of cosets $\{gG_0\}$, let $p : G \to X$ be the projection, let $x_0 = p(e)$. G acts on X by left multiplication. For any loop $w = \{g_t\}_{t\in I}$ in G at e, the cyclic homotopy $\{g_t \cdot\}_{t\in I} : \mathrm{id}_X \simeq \mathrm{id}_X : X \to X$ has trace $p \circ w$, hence $p_\pi \pi_1(G,e) \subset J(X,x_0)$. On the other hand, by Gleason's theorem (cf. [Steenrod] §7.5), $p : G \to X$ is a fiber bundle with fiber G_0. The exact homotopy sequence tells us that $p_\pi : \pi_1(G,e) \to \pi_1(X,x_0)$ is an epimorphism when G_0 is connected. □

The proof of the last case suggests an easy way to get information about $J(X)$ from the symmetry of the space X.

3.12 LEMMA. Let G be a topological group acting on X, i.e. a map $G \times X \xrightarrow{\cdot} X$ is given with the property $g' \cdot (g \cdot x) = (g'g) \cdot x$. Let $\xi : G \to X$ be the evaluation map at the base point x_0:

$$\xi : G,e \to X,x_0, \qquad g \mapsto g \cdot x_0 .$$

Then $\xi_\pi(\pi_1(G,e)) \subset J(X) = J(\mathrm{id},x_0)$. □

When the polyhedron X is aspherical (or, equivalently, the universal covering $\tilde{X}$ is contractible), the subgroup $J(f)$ can be given explicitly.

3.13 THEOREM. Let X be a connected aspherical (in the sense that $\pi_i(X) = 0$ for $i > 1$) polyhedron. Then for any $f : X \to X$,

$$J(f) = J(f,x_0) = Z(f_\pi \pi_1(X,x_0), \pi_1(X,f(x_0))) ,$$

or, equivalently, $J(f) = Z(\tilde{f}_\pi(\pi), \pi)$. In particular, $J(X) = Z(\pi)$.

PROOF. The proof is by obstruction theory. See [Brown (1971)] p. 102. □

Note that this theorem applies to compact surfaces with non-positive Euler characteristic.

4. PERMUTATIONS INDUCED BY CYCLIC HOMOTOPIES. We now put the idea introduced at the beginning of §3 to work.

4.1 THEOREM. Suppose $\tilde{f}(\pi) \subset J(\tilde{f})$. Then any two fixed point classes of f have the same index. If $L(f) = 0$ then $N(f) = 0$. If $L(f) \neq 0$ then $N(f) = R(f) = \#\mathrm{Coker}(1 - f_{1*})$.

In particular, when $J(\mathrm{id}_{\tilde{X}}) = \pi$, the above conclusion holds for any $f : X \to X$.

PROOF. For any $\alpha \in \pi$, we know $p\ \mathrm{Fix}(\alpha \circ \tilde{f}) = p\ \mathrm{Fix}(\tilde{f}_\pi(\alpha) \circ \tilde{f})$ by Lemma 1.4. Since $\tilde{f}_\pi(\alpha) \in J(\tilde{f})$, there is a homotopy $\{h_t\} : f \simeq f$ which lifts to $\{\tilde{h}_t\} : \tilde{f} \simeq \tilde{f}_\pi(\alpha) \circ \tilde{f}$. Hence, by the homotopy invariance theorem I.4.5

$$\mathrm{index}(f, p\ \mathrm{Fix}(\tilde{f})) = \mathrm{index}(f, p\ \mathrm{Fix}(\alpha \circ \tilde{f})) .$$

The α is arbitrary, so any two fixed point classes of f have the same index.

It immediately follows that $L(f) = 0$ implies $N(f) = 0$ and $L(f) \neq 0$ implies $N(f) = R(f)$. By Lemma 3.3, $\tilde{f}_\pi(\pi) \subset J(\tilde{f}) \subset Z(\tilde{f}_\pi(\pi), \pi)$, so $\tilde{f}_\pi(\pi)$ is abelian. Hence by Theorem 2.5, $R(f) = \#\mathrm{Coker}(1 - f_{1*})$.

In particular, when $J(\mathrm{id}_{\tilde{X}}) = \pi$, then for any $f : X \to X$ and any lifting $\tilde{f}$ we have $\tilde{f}_\pi(\pi) \subset J(\tilde{f}) = J(\mathrm{id}_{\tilde{X}}) = \pi$ by Lemma 3.4. □

By means of Lemma 3.6, we can restate this theorem as

4.2 THEOREM. Suppose $f_\pi \pi_1(X) \subset J(f)$. Then any two fixed point classes of f have the same index. If $L(f) = 0$ then $N(f) = 0$. If $L(f) \neq 0$ then $N(f) = R(f) = \#\mathrm{Coker}(1 - f_{1*})$.

In particular, when $J(X) = \pi_1(X)$ the above conclusion holds for any $f : X \to X$. □

EXAMPLE 1. $X = S^1$ and $\deg f = d$. Then $N(f) = |1 - d|$.

EXAMPLE 2. $X = T^n = S^1 \times \cdots \times S^1$, and f_{1*} is given by the integral matrix A. Then $N(f) = |\det(E - A)|$, where E is the identity matrix. Note that $L(f) = \det(E - A)$, and $\#\mathrm{Coker}(1 - f_{1*}) = |\det(E - A)|$ if $\det(E - A) \neq 0$, while $\#\mathrm{Coker}(1 - f_{1*}) = \infty$ if $\det(E - A) = 0$.

4.3 REMARK. The property $J(X) = \pi(X)$ is a homotopy type invariant of the space X.

The property $f_\pi(\pi) \subset J(f)$ is a homotopy type invariant of the map f.

4.4 COROLLARY. Let X be a polyhedron satisfying the condition $J(X) = \pi_1(X)$ (cf. 3.11). Then, for any map $f : X \to X$, $L(f) = 0$ implies $N(f) = 0$ while $L(f) \neq 0$ implies $N(f) = \#\mathrm{Coker}(1 - f_{1*})$. □

EXAMPLE 3. $X = L(m;q_1,\ldots,q_n)$, a generalized lens space, and $f : X \to X$ is of degree d. Then $N(f) = 0$ if $d = 1$ and $N(f) = (m,1 - d)$ if $d \neq 1$, where $(m,1 - d)$ is the greatest common divisor of m and $1 - d$.

Theorem 4.1 is about an extreme case. Turning to the general case, we now discuss the action of cyclic homotopies on the set of fixed point classes (or, equivalently, on the set of lifting classes).

Let $\tau \in J(\tilde{f})$. Let $\{h_t\} : f \simeq f$ be a cyclic homotopy which lifts to $\{\tilde{h}_t\} : \tilde{f} \simeq \tau \circ \tilde{f}$. Then, for any $\alpha \in \pi$, the homotopy $\{h_t\}$ also lifts to $\{\alpha \circ \tilde{h}_t\} : \alpha \circ \tilde{f} \simeq \alpha \circ (\tau \circ \tilde{f}) = \alpha\tau \circ \tilde{f}$, so that by I.2.3, the lifting class $[\alpha \circ \tilde{f}]$ corresponds to $[\alpha\tau \circ f]$ via $\{h_t\}$. This analysis motivates the following.

4.5 DEFINITION. Let $\tau \in J(\tilde{f})$. The action of τ on the set of liftings is $\tau : \alpha \circ \tilde{f} \mapsto \alpha\tau \circ \tilde{f}$; the action of τ on the set of lifting classes is $\tau : [\alpha \circ \tilde{f}] \mapsto [\alpha\tau \circ \tilde{f}]$; the action of τ on π is $\tau : \alpha \mapsto \alpha\tau$; the action of τ on the set π' of $\tilde{f}_\pi$-conjugacy classes is $\tau : [\alpha] \mapsto [\alpha\tau]$.

4.6 PROPOSITION. The above definition defines the action of $J(\tilde{f})$, from the right, on the respective sets. Two lifting classes (or fixed point classes) are permutable via cyclic homotopies iff they are in the same orbit of this action of $J(\tilde{f})$. □

4.7 NOTATION. Let $J_H(f) = \eta \circ \theta(J(\tilde{f})) = \eta \circ \theta(J(f))$, and $J_H(X) = \eta \circ \theta(J(X))$. Here $\eta \circ \theta$ is as in 2.1, the last equality is by 3.6.

4.8 PROPOSITION. If $\eta \circ \theta : \pi \to \mathrm{Coker}(1 - f_{1*})$ induces a 1-1 correspondence $\theta' : \pi' \to \mathrm{Coker}(1 - f_{1*})$, in particular, if f is eventually commutative (cf. 2.3-5), then the orbit length of the action of $J(\tilde{f})$ on π' is a constant $\# J_H(f)$.

PROOF. The following diagram is evidently commutative:

$$\begin{array}{ccc}
\pi' \times J(\tilde{f}) & \xrightarrow{([\alpha],\tau)\mapsto[\alpha\tau]} & \pi' \\
{\scriptstyle \theta'}\downarrow \quad \Vert & & \downarrow{\scriptstyle \theta'} \\
\mathrm{Coker}(1-f_{1*}) \times J(\tilde{f}) & \xrightarrow{(a,\tau)\mapsto a+\eta\circ\theta(\tau)} & \mathrm{Coker}(1-f_{1*}) \\
\Vert \quad \downarrow{\scriptstyle \eta\circ\theta} & & \Vert \\
\mathrm{Coker}(1-f_{1*}) \times J_H(f) & \xrightarrow{(a,t)\to a+t} & \mathrm{Coker}(1-f_{1*})
\end{array}$$

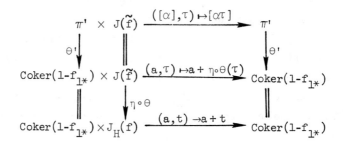

The horizontal arrows are group actions. The orbit structure of the first and the second are the same, since θ' is assumed to be 1-1. The orbit lengths of the second and the third actions are the same, since $\eta \circ \theta$ is an epimorphism. But the orbits of the third action are nothing but the cosets mod $J_H(f)$, so they are all of length $\# J_H(f)$.

Clearly the orbit structures of $J(\tilde{f})$ on π' and on the set of lifting classes are the same. The next theorem says that $\# J_H(f)$ is always a common divisor of all these orbit lengths.

4.9 THEOREM. Of all the orbit lengths of the action of $J(\tilde{f})$ on the set of fixed point classes, $\# J_H(f)$ is a common divisor. Hence $\# J_H(f)$ divides $N(f)$, $R(f)$ and $L(f)$.

PROOF. The isotropy subgroup at $[\alpha] \in \pi'$ of the action $J(\tilde{f})$ on π' is $S_\alpha = \{\tau \in J(\tilde{f}) \mid [\alpha\tau] = [\alpha]\}$. Hence the length of the orbit at $[\alpha]$ equals $[J(\tilde{f}) : S_\alpha]$, the index of the subgroup S_α in $J(\tilde{f})$. But $[\alpha\tau] = [\alpha]$ iff there is $\gamma \in \pi$ such that $\alpha\tau = \gamma\alpha\tilde{f}_\pi(\gamma^{-1})$, that is $\tau = \alpha^{-1}\gamma\alpha\tilde{f}_\pi(\gamma^{-1})$. Hence $\eta \circ \theta(S_\alpha) = 0$. In other words, S_α is contained in the kernel K of the epimorphism $\eta \circ \theta : J(\tilde{f}) \to J_H(f)$. Thus $\# J_H(f) = [J(\tilde{f}) : K]$ divides the orbit length $[J(\tilde{f}) : S_\alpha]$. This proves the first statement. The second statement then follows from 4.6 and the homotopy invariance theorem I.4.5. □

4.10 NOTATION. The divisibility symbol $m \mid n$, when m and n are allowed to be 0 or ∞, means one of the following:

(1) $n = 0$,

(2) $n = \infty$ and $m \neq 0$,

(3) both n and m are nonzero integers, and $m \mid n$ in the ordinary sense.

4.11 COROLLARY. Let the polyhedron X be a homogeneous space G/G_0 where G is a topological group, G_0 a compact Lie subgroup. Let k be the number of components of G_0. Let $f : X \to X$ be a map. Then, if $\#\mathrm{Coker}(1 - f_{1*}) = \infty$, we have $N(f) = 0$. If $\#\mathrm{Coker}(1 - f_{1*}) = h < \infty$, then the number $\frac{h}{(h,k)}$ divides $N(f)$, $R(f)$ and $L(f)$, where (h,k) is the greatest common divisor of h and k.

PROOF. Recall the proof of case (iv) in 3.11. From the exact homotopy sequence of the projection $p : G \to X$, we see the index $[\pi_1(X,x_0) : p_\pi\pi_1(G,e)] = k$, hence the index $[\pi_1(X) : J(X)] \mid k$. Therefore $[\mathrm{Coker}(1 - f_{1*}) : J_H(f)] \mid [\mathrm{Coker}(1 - f_{1*}) : J_H(X)] \mid k$. Now if $h = \infty$, then $\# J_H(f) = \infty$, by 4.9 we have $N(f) = 0$. If $h < \infty$, then $\frac{h}{(h,k)} \mid \# J_H(f)$. Apply 4.9. □

The results of this section permit us to compute $N(f)$ for quite a few spaces. But the following proposition indicates a serious limitation.

4.12 PROPOSITION. Let X be a connected compact polyhedron with nonzero Euler characteristic $\chi(X)$. Then $J(X)$ is trivial.

PROOF. A lifting of id_X is nothing but an element of π. Since by Lemma 3.3 $J(id_{\tilde{X}}) \subset Z(\pi)$, every element $\alpha \in J(id_{\tilde{X}})$ is a conjugacy class by itself. Suppose $\alpha \neq 1$. Then by the homotopy invariance I.4.5,

$$\text{index}(id_X, p\ \text{Fix}(id_{\tilde{X}})) = \text{index}(\alpha, p\ \text{Fix}(\alpha)) .$$

But the left hand side is $\chi(X)$, while the right hand side is 0 since $p\ \text{Fix}(\alpha)$ is empty. Thus $\chi(X) \neq 0$ implies $J(id_{\tilde{X}})$ is trivial. □

Applying this to an aspherical polyhedron, we get an interesting algebraic result.

4.13 COROLLARY. Suppose the group π is such that there is a compact $K(\pi,1)$ polyhedron, and $\chi(K(\pi,1)) \neq 0$. Then the center of π is trivial.

PROOF. Apply 4.12 and 3.13. □

5. POLYHEDRA WITH A FINITE FUNDAMENTAL GROUP. For a compact polyhedron X with finite fundamental group $\pi_1(X)$, the universal covering space $\tilde{X}$ is compact, so that we have the fixed point index as a tool also in $\tilde{X}$. In particular, we can explore the relation between $L(\tilde{f})$ and $\text{index}(p\ \text{Fix}(\tilde{f}))$.

Let $f : X \to X$ be a map, and let $\tilde{f} : \tilde{X} \to \tilde{X}$ be a lifting. Let $\tilde{x} \in \text{Fix}(\tilde{f})$ and $x = p(\tilde{x}) \in p\ \text{Fix}(\tilde{f})$. How many points are there in $\text{Fix}(\tilde{f}) \cap p^{-1}(x)$? Every point in $p^{-1}(x)$ has the form $\gamma\tilde{x}$, with $\gamma \in \pi$. Thus $\tilde{f}(\gamma\tilde{x}) = \gamma\tilde{x} \Leftrightarrow \tilde{f} \circ \gamma(\tilde{x}) = \gamma \circ \tilde{f}(\tilde{x}) \Leftrightarrow \tilde{f} \circ \gamma = \gamma \circ \tilde{f}$ (because both are liftings of f) $\Leftrightarrow \gamma \circ \tilde{f} = \tilde{f}_\pi(\gamma) \circ \tilde{f} \Leftrightarrow \gamma = \tilde{f}_\pi(\gamma) \Leftrightarrow \gamma \in \text{Fix}(\tilde{f}_\pi)$.

5.1 DEFINITION. The number $\mu([\tilde{f}])$ defined by $\mu([\tilde{f}]) = \#\text{Fix}(\tilde{f}_\pi)$, the order of the fixed-element-group $\text{Fix}(\tilde{f}_\pi)$, is called the <u>multiplicity</u> of the lifting class $[\tilde{f}]$, or of the fixed point class $p\ \text{Fix}(\tilde{f})$.

This number is independent of the choice of $\tilde{f}$ in $[\tilde{f}]$. In fact, another $\tilde{f}' \in [f]$ is of the form $\tilde{f}' = \delta \circ \tilde{f} \circ \delta^{-1}$, for some $\delta \in \pi$. Hence $\gamma \in \text{Fix}(\tilde{f}'_\pi) \Leftrightarrow \gamma = \tilde{f}'_\pi(\gamma) \Leftrightarrow \tilde{f}' \circ \gamma = \gamma \circ \tilde{f}' \Leftrightarrow \delta \circ \tilde{f} \circ \delta^{-1} \circ \gamma = \gamma \circ \delta \circ \tilde{f} \circ \delta^{-1}$ $\Leftrightarrow \tilde{f} \circ (\delta^{-1}\gamma\delta) = (\delta^{-1}\gamma\delta) \circ \tilde{f} \Leftrightarrow \delta^{-1}\gamma\delta = \tilde{f}_\pi(\delta^{-1}\gamma\delta) \Leftrightarrow \delta^{-1}\gamma\delta \in \text{Fix}(\tilde{f}_\pi)$, that is, $\text{Fix}(\tilde{f}') = \delta_*\text{Fix}(\tilde{f}_\pi)$, where δ_* is the inner automorphism $\gamma \mapsto \delta\gamma\delta^{-1}$.

5.2 LEMMA. $\mu([\tilde{f}]) \cdot \#[\tilde{f}] = \#\pi_1(X)$.

PROOF. The group π acts on $[\tilde{f}]$ by $\tilde{f} \overset{\gamma}{\mapsto} \gamma \circ \tilde{f} \circ \gamma^{-1}$. The action is transitive, and the isotropy subgroup is $\{\gamma \mid \gamma \circ \tilde{f} \circ \gamma^{-1} = \tilde{f}\} = \text{Fix}(\tilde{f}_\pi)$. □

5.3 THEOREM. $L(\tilde{f}) = \mu([\tilde{f}]) \cdot \text{index}(f, p\ \text{Fix}(\tilde{f}))$.

PROOF. By Hopf's Approximation Theorem (cf. [Brown (1971)] p. 118) there exists $g \simeq f$ with only isolated fixed points. Suppose this homotopy lifts to a homotopy $\tilde{g} \simeq \tilde{f}$. Then $\mu([\tilde{f}]) = \mu([\tilde{g}])$ since $\tilde{f}_\pi = \tilde{g}_\pi$, we have

$\text{index}(f, p\,\text{Fix}(\tilde{f})) = \text{index}(g, p\,\text{Fix}(\tilde{g}))$ by homotopy invariance, and $L(\tilde{f}) = L(\tilde{g})$. So, without loss of generality, we may assume that the map f in the theorem has only isolated fixed points.

Let $\tilde{x} \in \text{Fix}(\tilde{f})$, hence $x = p(\tilde{x}) \in p\,\text{Fix}(\tilde{f})$. The projection $p : \tilde{X} \to X$ is a local homeomorphism, and the fixed point index is a local invariant, so we have $\text{index}(\tilde{f}, \tilde{x}) = \text{index}(f, x)$. Every fixed point of f in $p\,\text{Fix}(\tilde{f})$ has $\mu([\tilde{f}])$ fixed points of $\tilde{f}$ above it. Hence

$$L(\tilde{f}) = \sum_{\tilde{x} \in \text{Fix}(\tilde{f})} \text{index}(\tilde{f}, \tilde{x}) = \mu([\tilde{f}]) \sum_{x \in p\text{Fix}(\tilde{f})} \text{index}(f, x)$$

$$= \mu([\tilde{f}]) \cdot \text{index}(f, p\,\text{Fix}(\tilde{f})) . \qquad \square$$

5.4 THEOREM (Averaging). $L(f) = \frac{1}{\#\pi} \sum_{\tilde{f}} L(\tilde{f})$, the summation is over all liftings of f.

PROOF.
$$\sum_{\tilde{f}} L(\tilde{f}) = \sum_{\tilde{f}} \mu([\tilde{f}])\text{index}(p\,\text{Fix}(\tilde{f})) \qquad \text{by 5.3}$$
$$= \sum_{[\tilde{f}]} \#[\tilde{f}] \cdot \mu([\tilde{f}]) \cdot \text{index}(p\,\text{Fix}(\tilde{f}))$$
$$= \sum_{[\tilde{f}]} (\#\pi) \cdot \text{index}(p\,\text{Fix}(\tilde{f})) \qquad \text{by 5.2}$$
$$= (\#\pi) \sum_{[\tilde{f}]} \text{index}(p\,\text{Fix}(\tilde{f}))$$
$$= (\#\pi) \cdot L(f) . \qquad \square$$

5.5 THEOREM. If $R(f) = \#\text{Coker}(1 - f_{1*})$ (in particular if f is eventually commutative), then $\mu([\tilde{f}]) = \#\text{Coker}(1 - f_{1*})$.

PROOF. We have $R(f) = \#\text{Coker}(1 - f_{1*})$ so by 2.3 the homomorphism $\eta \circ \theta : \pi \to \#\text{Coker}(1 - f_{1*})$ sends different $\tilde{f}_\pi$-conjugacy classes to different elements. Thus $\tilde{f}_\pi$-conjugacy classes are cosets in π, hence are all of the same cardinality. So all lifting classes of f consist of the same number (say k) of liftings, and $k \cdot \#\text{Coker}(1 - f_{1*}) = \#\pi$.

Now Lemma 5.2 tells us $k \cdot \mu([\tilde{f}]) = \#\pi$. Hence

$$\mu([\tilde{f}]) = \#\pi/k = \#\text{Coker}(1 - f_{1*}) . \qquad \square$$

Theorem 5.3 is very useful in the computation of $N(f)$. For example,

5.6 THEOREM. Let X be a connected compact polyhedron with finite fundamental group π. Suppose the action of π on the rational homology of the universal covering space $\tilde{X}$ is trivial, i.e. for every covering translation $\alpha \in \pi$, $\alpha_* = \text{id} : H_*(\tilde{X};Q) \to H_*(\tilde{X};Q)$. Then, for any map $f : X \to X$, $L(f) = 0$ implies $N(f) = 0$; $L(f) \neq 0$ implies $N(f) = R(f) \geq \#\text{Coker}(1 - f_{1*})$. If f is eventually commutative, then $R(f) = \#\text{Coker}(1 - f_{1*})$ and the index of the fixed point classes are all equal.

PROOF. Under the assumption on X, any two liftings $\tilde{f}$ and $\alpha \circ \tilde{f}$ induce the same homology homomorphism $H_*(\tilde{X};Q) \xrightarrow{\tilde{f}_*} H_*(\tilde{X};Q)$, hence the same $L(\tilde{f})$. Then we may apply Theorem 5.3 to show that any two fixed point classes are either both essential or both inessential. The last statement is from Theorem 5.5. □

5.7 LEMMA. Let X be a polyhedron with finite fundamental group π, and let $p : \tilde{X} \to X$ be its universal covering. Then, the action of π on the rational homology of $\tilde{X}$ is trivial iff $H_*(\tilde{X};Q) \cong H_*(X;Q)$.

PROOF. It is known (by means of the transfer, cf. [Bredon] p. 120) that the homology homomorphism $p : H_*(\tilde{X};Q) \to H_*(X;Q)$ sends the subspace

$$H_*(\tilde{X};Q)^{\pi} = \{h \in H_*(\tilde{X};Q) \mid \alpha_*(h) = h \text{ for all } \alpha \in \pi\}$$

isomorphically onto $H_*(X;Q)$. So $H_*(\tilde{X};Q)^{\pi} = H_*(\tilde{X};Q)$ iff $H_*(\tilde{X};Q) \cong H_*(X;Q)$. □

5.8 COROLLARY. Let $\tilde{X}$ be a compact 1-connected polyhedron which is a rational homology n-sphere, n odd. Let π be a finite group acting freely on $\tilde{X}$, and $X = \tilde{X}/\pi$. Then Theorem 5.6 applies.

PROOF. The projection $p : \tilde{X} \to X = \tilde{X}/\pi$ is a universal covering space of X. For every $\alpha \in \pi$, the degree of $\alpha : \tilde{X} \to \tilde{X}$ must be 1, because $L(\alpha) = 0$ (α has no fixed points). Hence $\alpha_* = 1 : H_*(\tilde{X};Q) \to H_*(\tilde{X};Q)$. □

Note that in Corollary 5.8 we restrict to n odd.

EXERCISE. Use Theorem 5.3 to discuss $N(f)$ for $\mathbb{R}P^{2k}$.

5.9 COROLLARY. If X is a closed 3-manifold with finite π_1, then Theorem 5.6 applies.

PROOF. $\tilde{X}$ is an (orientable) simply-connected 3-manifold, hence a homology 3-sphere. Apply Corollary 5.8. □

5.10 COROLLARY. Let π be a finite group acting freely on an odd dimensional sphere, and let X be the quotient space. Then for any $f : X \to X$, $L(f) = 0$ implies $N(f) = 0$, $L(f) \neq 0$ implies $N(f) = R(f)$. □

6. CONVERSES OF THE LEFSCHETZ FIXED POINT THEOREM. The celebrated Lefschetz fixed point theorem says that $L(f) \neq 0$ implies that every map homotopic to f has a fixed point, i.e. $MF[f] > 0$, where $MF[f]$ is as defined in §I.6. Its converse statement, "$L(f) = 0$ implies $MF[f] = 0$," is not always true even for homeomorphisms of closed manifolds, as shown by examples in [McCord]. It is desirable to understand under what restrictions on the space the converse does hold true.

The theory of fixed point classes provides the tools to attack this problem. The link between $L(f)$ and $MF[f]$ is $N(f)$.

Let X be a compact connected polyhedron. In this chapter we have given conditions for the implication "$L(f) = 0$ implies $N(f) = 0$," and in §I.6 we stated geometric conditions for the equality $N(f) = MF[f]$. Combining these two, we then get a converse of the Lefschetz theorem.

6.1 THEOREM. Let X be a compact connected polyhedron without global separating points. Suppose X satisfies the condition $J(X) = \pi_1(X)$. Then, for any map $f: X \to X$, $L(f) = 0$ iff f is homotopic to a fixed point free map.

PROOF. By the combination of Theorems I.6.3 and 4.4, we see that if X has no local separating points and is not a surface of negative Euler characteristic, and if $J(X) = \pi_1(X)$, then the conclusion is true. The condition "X is not a surface of negative Euler characteristic" can be deleted, since the other condition $J(X) = \pi_1(X)$ implies that $\pi_1(X)$ is abelian (Lemma 3.7). It remains to weaken the restriction about local separating points to that about global separating points.

So, suppose X has a local separating point which is not a global separating point. According to Lemma 6.4 below, X has S^1 as a strong deformation retract. Let $S^1 \xrightarrow{i} X \xrightarrow{r} S^1$ be the inclusion and retraction, and let $\varphi = r \circ f \circ i$. Then $N(f) = N(\varphi)$, by the homotopy type invariance I.5.4 of the Nielsen number. But for S^1 we know φ can be homotoped to a map ψ with exactly $N(\varphi)$ fixed points (if $\deg \varphi = n$, then take $\psi(z) = -z^n$). So, on X, the map $f \simeq i \circ \varphi \circ r$ can be homotoped to $i \circ \psi \circ r$ with exactly $N(f)$ fixed points, i.e. $N(f) = MF[f]$. □

6.2 THEOREM. Let X be a compact connected polyhedron without global separating points. Suppose $\pi_1(X)$ is finite and the universal covering space $\tilde{X}$ has the same rational homology as X. Then, for any map $f: X \to X$, $L(f) = 0$ iff f is homotopic to a fixed point free map.

PROOF. Since $\pi_1(X)$ is finite, X cannot be a surface of negative Euler characteristic. And X cannot have a local separating point which is not a global one, by the first part of Lemma 6.4 below. So, according to Theorem I.6.3, $N(f) = MF[f]$ for any map $f: X \to X$.

On the other hand, Theorem 5.6 and Lemma 5.7 tell us that $L(f) = 0$ implies $N(f) = 0$. Hence the conclusion. □

A common special case of both theorems is:

6.3 COROLLARY. Let X be a compact connected and simply-connected polyhedron without global separating points. Then, for a map $f: X \to X$, $L(f) = 0$ iff f is homotopic to a fixed point free map. □

The restriction on global separating points cannot be removed. See the discussion in [Brown (1971)] §VIII.F.

The lemma used in the above proofs follows.

6.4 LEMMA. If a connected polyhedron X has a local separating point x which is not a global separating point, then X has a retract homeomorphic to S^1. If, in addition, $J(X) = \pi_1(X)$, then X has a strong deformation retract homeomorphic to S^1.

PROOF. By cutting X at x we see that X is obtained from a connected polyhedron Y by identifying two points $y, y' \in Y$. There exists a piecewise linear arc A in Y from y to y'. According to Tietze's extension theorem, there is a retraction $r : Y \to A$. Hence X has a retraction onto the subspace C obtained from A by identifying y and y'. Obviously $C \cong S^1$.

Now add the assumption $J(X) = \pi_1(X)$. We want to show Y has A as a strong deformation retract, hence X has C as a strong deformation retract. It is well known that a subpolyhedron is a strong deformation retract iff the inclusion map is a homotopy equivalence, so it remains to show that Y is contractible.

Construct an infinite cyclic covering space $\tilde{X}$ of X as follows. Take copies Y_i of Y, $i \in \mathbb{Z}$, and identify y_i' with y_{i+1} for every i, the result is $\tilde{X}$. The projection $p : \tilde{X} \to X$ is defined naturally via the identification from Y to X. A generator $\alpha : \tilde{X} \to \tilde{X}$ of the group of covering translations sends Y_i onto Y_{i+1}.

Since $\tilde{X} = \cdots \vee Y_{-1} \vee Y_0 \vee Y_1 \vee \cdots$, $\pi_1(\tilde{X})$ should be isomorphic to the free product of all those $\pi_1(Y_i)$, $i \in \mathbb{Z}$. But the condition $J(X) = \pi_1(X)$ implies that $\pi_1(X)$ is abelian, hence $\pi_1(\tilde{X})$ is also abelian. This is possible only if the free product is trivial. Hence $\pi_1(Y)$ is trivial, and $\tilde{X}$ is the universal covering of X.

Now look at $\tilde{H}_*(\tilde{X}) = \Sigma_i \tilde{H}_*(Y_i)$. The homomorphism $\alpha_* : \tilde{H}_*(\tilde{X}) \to \tilde{H}_*(\tilde{X})$ sends $\tilde{H}_*(Y_i)$ onto $\tilde{H}_*(Y_{i+1})$. But $J(X) = \pi_1(X)$ implies there exists a cyclic homotopy $\mathrm{id} \simeq \mathrm{id} : X \to X$ which lifts to a homotopy $\mathrm{id} \simeq \alpha : \tilde{X} \to \tilde{X}$, hence $\alpha_* = 1 : \tilde{H}_*(\tilde{X}) \to \tilde{H}_*(\tilde{X})$. It follows that $\tilde{H}_*(Y) = 0$. Thus the polyhedron Y is simply-connected and acyclic, by the Whitehead Theorem (cf. [Spanier] pp. 405-406) it is contractible. □

CHAPTER III

THE FIXED POINT CLASS FUNCTOR

The language of categories and functors is often very convenient for describing structural aspects of a theory. In the theory of fixed point classes, there are three hierarchies of concern, that of subspaces, that of regular covering spaces, and that of iterates of the self-map. In other words, we are interested in the relationship between the fixed point classes of a map and those of its restrictions on subspaces, the relationship between the ordinary fixed point classes and the coarser classes defined via (non-universal) regular covering spaces, and the relationship between fixed point classes of a map f and those of iterates of f. We will discuss these three aspects in §§1-3 respectively. Sections 1-2 are prerequisite to Chapter IV. Section 4, being a sequel of §3, is an introduction to the Nielsen type theory of periodic points.

Our use of the language of categories and functors is at a rather naïve level. It is not clear whether the theory of fixed point classes admits a deep categorical study such as what Dold has done in [Dold (1974)].

1. THE CATEGORY OF SELF-MAPS AND THE FIXED-POINT-CLASS FUNCTOR. In this section we introduce the language of category and functor into fixed point theory, then prove the commutativity theorem as promised at the end of Chapter I. We also give a few general facts about liftings for future reference.

The spaces considered in this section are assumed to have universal covering spaces. The universal covering space of X will be denoted $p_X : \tilde{X} \to X$, or simply $p : \tilde{X} \to X$. As usual (cf. II.1.2), we will identify $\pi_1(X)$ with the group $\mathcal{D}$ of covering translations.

1.1 DEFINITION. Let $h : X \to Y$ be a map. We say a map $\tilde{h} : \tilde{X} \to \tilde{Y}$ is a <u>lifting</u> of h, or h lifts to $\tilde{h}$, if $p_Y \circ \tilde{h} = h \circ p_X$, i.e. if the diagram

$$\begin{array}{ccc} \tilde{X} & \xrightarrow{\tilde{h}} & \tilde{Y} \\ \downarrow{p_X} & & \downarrow{p_Y} \\ X & \xrightarrow{h} & Y \end{array}$$

commutes.

The next two propositions are easy facts from covering space theory.

1.2 PROPOSITION. (i) Suppose $x_0 \in X$ and $y_0 = h(x_0)$; also $\tilde{x}_0 \in p_X^{-1}(x_0)$ and $\tilde{y}_0 \in p_Y^{-1}(y_0)$. Then there is a unique lifting $\tilde{h}$ of h such that $\tilde{h}(\tilde{x}_0) = \tilde{y}_0$.

(ii) For any two liftings $\tilde{h}$ and $\tilde{h}'$ of h, there is a unique $\beta \in \pi_1(Y)$ such that $\tilde{h}' = \beta \circ \tilde{h}$. □

1.3 PROPOSITION. Let $X \xrightarrow{f} Y \xrightarrow{g} Z$ be maps.

(i) A map $\tilde{X} \to \tilde{Z}$ is a lifting of $g \circ f$ iff it can be factored as $\tilde{g} \circ \tilde{f}$, where $\tilde{f}$ and $\tilde{g}$ are liftings of f and g respectively.

(ii) If $\tilde{g} \circ \tilde{f} = \tilde{g}' \circ \tilde{f}'$, then there exists $\beta \in \pi_1(Y)$ such that $\tilde{f}' = \beta \circ \tilde{f}$ and $\tilde{g}' = \tilde{g} \circ \beta^{-1}$. □

1.4 DEFINITION. Let $f: X \to X$ and $g: Y \to Y$ be two self-maps. A morphism from f to g is a map $h: X \to Y$ such that $h \circ f = g \circ h$, i.e. the diagram

$$\begin{array}{ccc} X & \xrightarrow{f} & X \\ {\scriptstyle h}\downarrow & & \downarrow{\scriptstyle h} \\ Y & \xrightarrow{g} & Y \end{array}$$

commutes. Self-maps and morphisms between them form a category -- the category of self-maps.

1.5 PROPOSITION. Let

$$\begin{array}{ccc} X & \xrightarrow{f} & X \\ {\scriptstyle h}\downarrow & & \downarrow{\scriptstyle h} \\ Y & \xrightarrow{g} & Y \end{array}$$

be a morphism of self-maps. Given a lifting $\tilde{f}$ of f and a lifting $\tilde{h}$ of h, there is a unique lifting $\tilde{g}$ of g such that

$$\begin{array}{ccc} \tilde{X} & \xrightarrow{\tilde{f}} & \tilde{X} \\ {\scriptstyle \tilde{h}}\downarrow & & \downarrow{\scriptstyle \tilde{h}} \\ \tilde{Y} & \xrightarrow{\tilde{g}} & \tilde{Y} \end{array}$$

commutes.

PROOF. Pick a point $\tilde{x}_0 \in \tilde{X}$, and let $\tilde{y}_0 = \tilde{h}(\tilde{x}_0) \in \tilde{Y}$. There is a unique lifting $\tilde{g}$ of g such that $\tilde{g}(\tilde{y}_0) = \tilde{h}(\tilde{f}(\tilde{x}_0))$. Now $\tilde{h} \circ \tilde{f}$ and $\tilde{g} \circ \tilde{h} : \tilde{X} \to \tilde{Y}$ are both liftings of the same map $h \circ f = g \circ h : X \to Y$, and they agree at $\tilde{x}_0$. By the unique lifting property of covering spaces, we have $\tilde{h} \circ \tilde{f} = \tilde{g} \circ \tilde{h}$. □

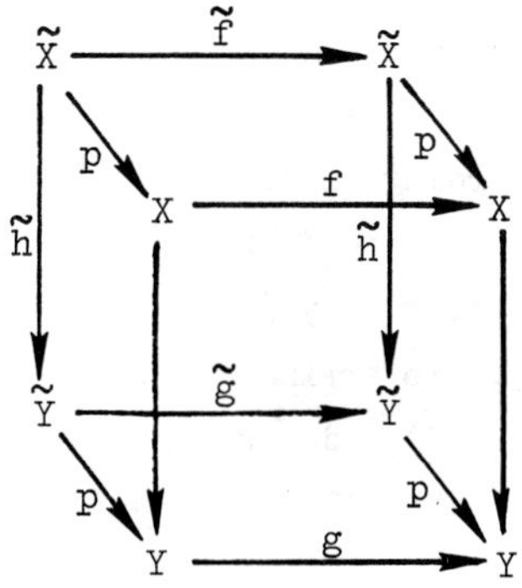

1.6 NOTATION. A lifting $\tilde{h}$ of h determines a correspondence $\tilde{h}_{lift}$ from liftings of f to liftings of g by the above proposition. Thus $\tilde{h} \circ \tilde{f} = \tilde{h}_{lift}(\tilde{f}) \circ \tilde{h}$. In particular, consider the morphism

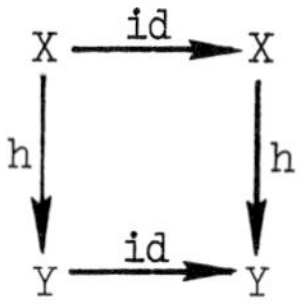

,

then $\tilde{h}$ determines a correspondence from $\pi_1(X)$ to $\pi_1(Y)$, both identified with the group of covering translations. This correspondence will be denoted $\tilde{h}_\pi : \pi_1(X) \to \pi_1(Y)$, it is characterized by the formula $\tilde{h} \circ \alpha = \tilde{h}_\pi(\alpha) \circ \tilde{h}$.

1.7 PROPOSITION. (i) $(\tilde{h}_2 \circ \tilde{h}_1)_{lift} = \tilde{h}_{2\,lift} \circ \tilde{h}_{1\,lift}$.

(ii) $\tilde{h}_{lift}(\tilde{f}_2 \circ \tilde{f}_1) = \tilde{h}_{lift}(\tilde{f}_2) \circ \tilde{h}_{lift}(\tilde{f}_1)$.

(iii) $\tilde{h}_\pi$ is a homomorphism. □

1.8 THEOREM. (i) The correspondence $\tilde{h}_{lift}$ sends a lifting class of f into a lifting class of g, that is, $\tilde{h}_{lift}([\tilde{f}]) \subset [\tilde{h}_{lift}(\tilde{f})]$.

(ii) On the lifting class level, this correspondence does not depend on the lifting $\tilde{h}$ but is determined by h itself. We will write $h_{FPC} : [\tilde{f}] \mapsto [\tilde{g}]$ henceforth. And we have $(h_2 \circ h_1)_{FPC} = h_{2\,FPC} \circ h_{1\,FPC}$.

(iii) Every fixed point class of f is mapped by h into some fixed point class of g. Namely, if $h_{FPC} : [\tilde{f}] \mapsto [\tilde{g}]$, then $h(p_X\text{Fix}(\tilde{f})) \subset p_Y\text{Fix}(\tilde{g})$.

PROOF. (i) If $\tilde{f}' = \alpha \circ \tilde{f} \circ \alpha^{-1}$, where $\alpha \in \pi_1(X)$, and $\tilde{g} = \tilde{h}_{lift}(\tilde{f})$, then $\tilde{h}_{lift}(\tilde{f}') = \tilde{h}_\pi(\alpha) \circ \tilde{g} \circ \tilde{h}_\pi(\alpha^{-1})$.

(ii) If $\tilde{h}' = \beta \circ \tilde{h}$ is another lifting of h, where $\beta \in \pi_1(Y)$, and $\tilde{h}_{lift} : \tilde{f} \mapsto \tilde{g}$, then $\tilde{h}'_{lift} : \tilde{f} \mapsto \beta \circ \tilde{g} \circ \beta^{-1}$. Apply 1.7 (ii) to get the second conclusion.

(iii) If $\tilde{h}_{lift} : \tilde{f} \mapsto \tilde{g}$, then obviously $\tilde{h}\,\text{Fix}(\tilde{f}) \subset \text{Fix}(\tilde{g})$. □

1.9 DEFINITION. When (and only when) $h_{FPC} : [\tilde{f}] \mapsto [\tilde{g}]$, we say that h <u>maps the fixed point class</u> $p_X \mathrm{Fix}(\tilde{f})$ <u>into the fixed point class</u> $p_Y \mathrm{Fix}(\tilde{g})$.

Recall that a fixed point class is always labelled by a lifting class. According to the above definition, every fixed point class of f is mapped by h into a <u>unique</u> fixed point class of g, even when the class $p\,\mathrm{Fix}(\tilde{f})$ is empty. By an abuse of language, we will speak of the fixed point class $[\tilde{f}]$, referring either to the lifting class $[\tilde{f}]$ or to the point set $p\,\mathrm{Fix}(\tilde{f})$, depending on the context.

EXERCISE. For a lifting class $[\tilde{g}]$ of g, we have

$$\mathrm{Fix}(f) \cap h^{-1} p_Y \mathrm{Fix}(\tilde{g}) = \bigcup_{[\tilde{f}] \in h_{FPC}^{-1}[\tilde{g}]} p_X \mathrm{Fix}(\tilde{f}) .$$

1.10 DEFINITION. Let X be a compact connected polyhedron, $f : X \to X$ be a map. The <u>fixed-point-class data</u> of f, denoted $\mathrm{FPC}(f)$, is the weighted set of conjugacy classes of liftings $\tilde{f} : \tilde{X} \to \tilde{X}$ of f, the weight of a class $[\tilde{f}]$ being $\mathrm{index}(f, p\,\mathrm{Fix}(\tilde{f}))$. The rule that assigns $\mathrm{FPC}(f)$ to a self-map f and assigns the correspondence $h_{FPC} : \mathrm{FPC}(f) \to \mathrm{FPC}(g)$ of Theorem 1.8 (ii) to a morphism

$$\begin{array}{ccc} X & \xrightarrow{f} & X \\ {\scriptstyle h}\downarrow & & \downarrow{\scriptstyle h} \\ Y & \xrightarrow{g} & Y \end{array}$$

is a covariant functor from the category of self-maps of compact connected polyhedra to the category of weighted sets. This functor will be referred to as the <u>fixed-point-class functor</u>. Note that the morphisms in the latter category are just correspondences of sets, with no restriction on their behavior with respect to the weight.

Let us use this language to give a proof of the commutativity theorem.

1.11 LEMMA. Consider the morphism

$$\begin{array}{ccc} X & \xrightarrow{f} & X \\ {\scriptstyle f}\downarrow & & \downarrow{\scriptstyle f} \\ X & \xrightarrow[f]{} & X \end{array}$$

Then $f_{FPC} = \mathrm{id} : \mathrm{FPC}(f) \to \mathrm{FPC}(f)$.

PROOF. For any lifting $\tilde{f} : \tilde{X} \to \tilde{X}$ of f, we have a commutative diagram

$$\begin{array}{ccc} \tilde{X} & \xrightarrow{\tilde{f}} & \tilde{X} \\ {\scriptstyle \tilde{f}}\downarrow & & \downarrow{\scriptstyle \tilde{f}} \\ \tilde{X} & \xrightarrow{\tilde{f}} & \tilde{X} \end{array}$$

Hence $\tilde{f}_{lift}(\tilde{f}) = \tilde{f}$ by 1.6, and $f_{FPC} : [\tilde{f}] \mapsto [\tilde{f}]$ by 1.8 and 1.10. □

1.12 THEOREM (The commutativity of fixed-point-class data). Let X and Y be compact connected polyhedra, and let $f : X \to Y$ and $g : Y \to X$ be maps. Then $f_{FPC} : FPC(g \circ f) \to FPC(f \circ g)$ and $g_{FPC} : FPC(f \circ g) \to FPC(g \circ f)$ are a pair of index-preserving bijections between $FPC(g \circ f)$ and $FPC(f \circ g)$.

PROOF. Obviously we have commutative diagrams

$$\begin{array}{ccc} X & \xrightarrow{g\circ f} & X \\ {\scriptstyle f}\downarrow & & \downarrow{\scriptstyle f} \\ Y & \xrightarrow{f\circ g} & Y \end{array} \qquad \begin{array}{ccc} Y & \xrightarrow{f\circ g} & Y \\ {\scriptstyle g}\downarrow & & \downarrow{\scriptstyle g} \\ X & \xrightarrow{g\circ f} & X \end{array} ,$$

so f_{FPC} and g_{FPC} are defined. By the functorial property 1.8 (ii) and Lemma 1.11, $g_{FPC} \circ f_{FPC} = (g \circ f)_{FPC} = id : FPC(g \circ f) \to FPC(g \circ f)$, similarly $f_{FPC} \circ g_{FPC} = id : FPC(f \circ g) \to FPC(f \circ g)$. Thus f_{FPC} and g_{FPC} are a pair of bijections. It remains to show they preserve the index.

By 1.3, each lifting of $g \circ f$ is of the form $\tilde{g} \circ \tilde{f}$. The correspondence f_{FPC} sends $[\tilde{g} \circ \tilde{f}]$ to $[\tilde{f} \circ \tilde{g}]$ by Definition 1.7, since $\tilde{f} \circ (\tilde{g} \circ \tilde{f}) = (\tilde{f} \circ \tilde{g}) \circ \tilde{f}$. Evidently $p_X \mathrm{Fix}(\tilde{g} \circ \tilde{f}) \underset{g}{\overset{f}{\rightleftarrows}} p_Y \mathrm{Fix}(\tilde{f} \circ \tilde{g})$ is a pair of homeomorphisms. If both fixed point classes are empty, they have the same index 0; if both are non-empty, they also have the same index by Theorem I.5.2. Thus f_{FPC} is index-preserving. □

Now let us clarify some points in the above theory. The relationship between $\tilde{h}_\pi$ and h_π is similar to Lemma II.1.3.

1.13 LEMMA. Suppose $x_0 \in X$ and $\tilde{x}_0 \in p_X^{-1}(x_0)$ are the base points used in the identification II.1.2 of $\pi_1(X)$ with the group of covering translations of $p_X : \tilde{X} \to X$. And similarly $y_0 \in Y$ and $\tilde{y}_0 \in p_Y^{-1}(y_0)$. Let $h : X \to Y$ be a map, $\tilde{h} : \tilde{X} \to \tilde{Y}$ be a lifting. Let $\tilde{w}$ be a path in $\tilde{Y}$ from $\tilde{y}_0$ to $\tilde{h}(\tilde{x}_0)$, and $w = p_Y \circ \tilde{w}$. Then the following diagram commutes:

$$\pi_1(X,x_0) \xrightarrow{h_\pi} \pi_1(Y,h(x_0)) \xrightarrow{w_*} \pi_1(Y,y_0), \qquad \tilde{h}_\pi : \pi_1(X,x_0) \to \pi_1(Y,y_0)$$

In particular, if $y_0 = h(x_0)$ and $\tilde{y}_0 = \tilde{h}(\tilde{x}_0)$, then $\tilde{h}_\pi = h_\pi$. □

1.14 PROPOSITION. Let

$$\begin{array}{ccc} X & \xrightarrow{f} & X \\ {\scriptstyle h}\downarrow & & \downarrow{\scriptstyle h} \\ Y & \xrightarrow{g} & Y \end{array}$$

be a morphism of self-maps.

(i) If $h_\pi : \pi_1(X) \to \pi_1(Y)$ is surjective, then $\tilde{h}_{lift}$ is surjective, hence $h_{FPC} : FPC(f) \to FPC(g)$ is also surjective.

(ii) If $h_\pi : \pi_1(X) \to \pi_1(Y)$ is injective, then $\tilde{h}_{lift}$ is injective.

(iii) Suppose $h_{FPC}[\tilde{f}] = [\tilde{g}]$. Then there is a lifting $\tilde{h}$ of h such that $\tilde{h}_{lift}(\tilde{f}) = \tilde{g}$. □

Generally speaking, in (ii) above h_{FPC} is not injective. It is an intricate question to determine the cardinality of $h_{FPC}^{-1}[\tilde{g}]$, i.e. the number of fixed point classes of f that are mapped by h into a given fixed point class $[\tilde{g}]$. Note that the answer may vary with $[\tilde{g}]$.

1.15 PROPOSITION. Suppose the diagrams

$$\begin{array}{ccc} X & \xrightarrow{f} & X \\ {\scriptstyle h}\downarrow & & \downarrow{\scriptstyle h} \\ Y & \xrightarrow{g} & Y \end{array} \qquad \begin{array}{ccc} \tilde{X} & \xrightarrow{\tilde{f}} & \tilde{X} \\ {\scriptstyle \tilde{h}}\downarrow & & \downarrow{\scriptstyle \tilde{h}} \\ \tilde{Y} & \xrightarrow{\tilde{g}} & \tilde{Y} \end{array}$$

are commutative. Suppose $h_\pi : \pi_1(X) \to \pi_1(Y)$ is injective and $h_\pi \pi_1(X)$ is a normal subgroup of $\pi_1(Y)$. Then $\# h_{FPC}^{-1}[\tilde{g}] = [\mathrm{Fix}(\tilde{g}_*) : \xi\, \mathrm{Fix}(\tilde{g}_\pi)]$, where $\xi : \pi_1(Y) \to \pi_1(Y)/h_\pi\pi_1(X) = \mathrm{Coker}\, h_\pi$ is the projection and $\tilde{g}_* : \mathrm{Coker}\, h_\pi \to \mathrm{Coker}\, h_\pi$ is induced by $\tilde{g}_\pi : \pi_1(Y) \to \pi_1(Y)$.

PROOF. It is easy to check that the normal subgroup $\tilde{h}_\pi\pi_1(X) = h_\pi\pi_1(X)$ is invariant under $\tilde{g}_\pi$, so that the statement makes sense.

Suppose $\tilde{h}_{lift}$ sends a lifting $\alpha \circ \tilde{f}$, where $\alpha \in \pi_1(X)$, to a lifting $\beta^{-1} \circ \tilde{g} \circ \beta$ conjugate to $\tilde{g}$. This is equivalent to saying that $\tilde{h} \circ \alpha \circ \tilde{f} = \beta^{-1} \circ \tilde{g} \circ \beta \circ \tilde{h}$, or $\tilde{g}_\pi(\beta) = \beta \tilde{h}_\pi(\alpha)$. So, $\beta^{-1} \circ \tilde{g} \circ \beta$ is in the image of $\tilde{h}_{lift}$ iff $\xi(\beta) \in \mathrm{Fix}(\tilde{g}_*)$.

Now suppose $\tilde{f}_i$, $i = 1,2$, are two liftings of f, and $\tilde{h}_{lift}(f_i) = \beta_i^{-1} \circ \tilde{g} \circ \beta_i$. Then $[\tilde{f}_1] = [\tilde{f}_2]$ iff $\tilde{f}_2 = \gamma \circ \tilde{f}_1 \circ \gamma^{-1}$ for some $\gamma \in \pi_1(X)$. But by 1.6 (ii) and 1.14 (ii) we know $\tilde{f}_2 = \gamma \circ \tilde{f}_1 \circ \gamma^{-1}$ iff $\tilde{h}_{lift}(\tilde{f}_2) = \tilde{h}_\pi(\gamma) \circ \tilde{h}_{lift}(\tilde{f}_1) \circ \tilde{h}_\pi(\gamma^{-1})$. The latter equality is equivalent to $\beta_2^{-1} \circ \tilde{g} \circ \beta_2 = \tilde{h}_\pi(\gamma) \circ \beta_1^{-1} \circ \tilde{g} \circ \beta_1 \circ \tilde{h}_\pi(\gamma^{-1})$, or $\beta_2 \tilde{h}_\pi(\gamma) \beta_1^{-1} \in \mathrm{Fix}(\tilde{g}_\pi)$. So,

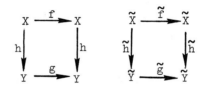

$[\tilde{f}_1] = [\tilde{f}_2]$ iff there exist $\bar{\gamma} \in h_\pi \pi_1(X)$ and $\delta \in \mathrm{Fix}(\tilde{g}_\pi)$ such that $\beta_2 \bar{\gamma} \beta_1^{-1} = \delta$, or $\delta\beta_1 = \beta_2\bar{\gamma}$; in other words, iff $\xi(\delta\beta_1) = \xi(\beta_2)$ for some $\delta \in \mathrm{Fix}(\tilde{g}_\pi)$; or, equivalently, $\xi(\beta_1)$ and $\xi(\beta_2)$ are in the same right coset of $\mathrm{Fix}(\tilde{g}_*)$ mod $\xi\,\mathrm{Fix}(\tilde{g}_\pi)$.

Hence $\# h_{\mathrm{FPC}}^{-1}[\tilde{g}]$ equals the index of $\xi\,\mathrm{Fix}(\tilde{g}_\pi)$ in $\mathrm{Fix}(\tilde{g}_*)$. □

We now turn to the homotopy behavior of the fixed-point-class functor. Let $\{f_t\}_{t\in I} : X \to X$ be a homotopy, and let $\mathbb{F} : X \times I \to X \times I$, $(x,t) \mapsto (f_t(x),t)$ be the fat homotopy (see Definition I.2.5). Let $i_t : X \to X \times I$, $x \mapsto (x,t)$ be the embedding. Then we have a morphism of self-maps

$$\begin{array}{ccc} X & \xrightarrow{f_t} & X \\ {\scriptstyle i_t}\downarrow & & \downarrow{\scriptstyle i_t} \\ X\times I & \xrightarrow{\mathbb{F}} & X\times I \end{array}$$

for every $t \in I$. The homotopy invariance theorem I.4.5 told us that $(i_t)_{\mathrm{FPC}} : \mathrm{FPC}(f_t) \to \mathrm{FPC}(\mathbb{F})$ is an index-preserving bijection, thus $(i_1)_{\mathrm{FPC}}^{-1} \circ (i_0)_{\mathrm{FPC}} : \mathrm{FPC}(f_0) \to \mathrm{FPC}(f_1)$ is also an index-preserving bijection. We call it the correspondence via the homotopy $\{f_t\}$ (cf. Definition I.2.3), and write $\{f_t\} : [\tilde{f}_0] \leadsto [\tilde{f}_1]$.

1.16 PROPOSITION. Let $\{f_t\}_{t\in I} : X \to X$ and $\{g_t\}_{t\in I} : Y \to Y$ be two homotopies of self-maps, and let $\{h_t\}_{t\in I} : X \to Y$ be a homotopy, such that $h_t \circ f_t = g_t \circ h_t$ for all $t \in I$. Let $\{\tilde{f}_t\}_{t\in I}$ be a lifting of $\{f_t\}_{t\in I}$, and $\{\tilde{h}_t\}_{t\in I}$ be a lifting of $\{h_t\}_{t\in I}$. Then there is a unique lifting $\{\tilde{g}_t\}_{t\in I}$ of $\{g_t\}_{t\in I}$ such that $\tilde{h}_t \circ \tilde{f}_t = \tilde{g}_t \circ \tilde{h}_t$ for all $t \in I$. Hence the diagram

$$\begin{array}{ccc} [\tilde{f}_0] & \overset{\{f_t\}}{\leadsto} & [\tilde{f}_1] \\ {\scriptstyle (h_0)_{\mathrm{FPC}}}\downarrow & & \downarrow{\scriptstyle (h_1)_{\mathrm{FPC}}} \\ [\tilde{g}_0] & \overset{\{g_t\}}{\leadsto} & [\tilde{g}_1] \end{array}$$

commutes.

In particular, $\tilde{h}_{0\pi} = \tilde{h}_{1\pi} : \pi_1(X) \to \pi_1(Y)$. □

Analogous to Definition II.3.1, we have:

1.17 DEFINITION. Let $h : X \to Y$ be a map, and let $\tilde{h} : \tilde{X} \to \tilde{Y}$ be a lifting. Define

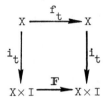

$$J(\tilde{h}) := \{\beta \in \pi_1(Y) \mid \text{there exists a cyclic homotopy } \{h_t\} : h \simeq h \text{ which lifts to } \{\tilde{h}_t\} : \tilde{h} \simeq \beta \circ \tilde{h}\} .$$

$J(\tilde{h})$ is a subgroup of $\pi_1(Y)$. Cf. Proposition II.3.2.

The proof of the next lemma is analogous to that of II.3.3-4.

1.18 LEMMA. Let $X \xrightarrow{h} Y \xrightarrow{k} Z$ be maps; let $\tilde{X} \xrightarrow{\tilde{h}} \tilde{Y} \xrightarrow{\tilde{k}} \tilde{Z}$ be their liftings.

(i) Elements of $J(\tilde{h})$ commute with elements of $\tilde{h}_\pi\pi_1(X)$, that is,

$$J(\tilde{h}) \subset \text{Centralizer of } \tilde{h}_\pi\pi_1(X) \text{ in } \pi_1(Y) .$$

(ii) $J(\tilde{k}) \subset J(\tilde{k} \circ \tilde{h})$ and $\tilde{k}_\pi J(\tilde{h}) \subset J(\tilde{k} \circ \tilde{h})$. □

2. FIXED POINT CLASSES MODULO A NORMAL SUBGROUP. In the theory of (ordinary) fixed point classes, we work on the universal covering space. The group of covering translations plays a key role, that is why the fundamental group comes in. It is not surprising that this theory can be generalized to work on all regular covering spaces.

Let $p' : X' \to X$ be a regular covering space of X. For $x_0 \in X$ and $x_0' \in p'^{-1}(x_0)$, the subgroup $K(x_0) = p'_\pi\pi_1(X', x_0')$ is a normal subgroup of $\pi_1(X, x_0)$. For any path w from x_0 to x_1, $w_* : \pi_1(X, x_1) \to \pi_1(X, x_0)$ sends $K(x_1)$ onto $K(x_0)$. Hence, as in the case of the universal covering space, the base point is not of much concern. We usually say K is a normal subgroup of $\pi_1(X)$ and write $K \lhd \pi_1(X)$, when we are not paying special attention to the base point. Regular covering spaces are, up to isomorphism, in 1-1 correspondence with normal subgroups of $\pi_1(X)$, the universal covering space corresponding to the trivial subgroup $\{1\}$. Given $K \lhd \pi_1(X)$, with $\pi_1(X)$ identified (cf. II.1.2) with the group of covering translations on the universal covering $p : \tilde{X} \to X$, we may consider the quotient space $\tilde{X}/K$ and obtain a commutative triangle of covering maps

$$\begin{array}{ccc} \tilde{X} & \xrightarrow{q_K} & \tilde{X}/K \\ & {\scriptstyle p}\searrow \quad \swarrow{\scriptstyle p_K} & \\ & X & \end{array} ,$$

and we will take $p_K : \tilde{X}/K \to X$ as the model of the coverings corresponding to K. The group $\mathcal{D}_K$ of covering translations on this regular covering space is the quotient group $\pi_1(X)/K$.

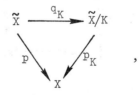

The only obstacle to developing a theory of fixed point classes with respect to a regular covering is that not every map $f : X \to X$ can be lifted to $\tilde{f}_K : \tilde{X}/K \to \tilde{X}/K$. We know from covering space theory that such a lifting exists iff $f_\pi(K) \subset K$, or, more precisely, $f_\pi(K(x)) \subset K(f(x))$ for some (hence every, since $K \lhd \pi_1(X)$) point $x \in X$.

So, in this section, <u>we restrict our attention to maps</u> f <u>with</u> $f_\pi K \subset K$. This property is clearly preserved under homotopy. If K is a fully invariant subgroup of $\pi_1(X)$ (in the sense that every endomorphism sends K into K), such as the commutator subgroup, then this is no restriction on f at all.

As we have seen in §I.1, the notion of fixed point class rests upon the existence of liftings (cf. Proposition I.1.2). So, under our present assumption, we can develop a similar theory by simply replacing $\tilde{X}$ and $\pi_1(X)$ with $\tilde{X}/K$ and $\pi_1(X)/K$ in every definition, every theorem and every proof, since everything was in terms of liftings and covering translations. What follows is a sample list of definitions and basic facts.

2.1 DEFINITION. A map $\tilde{f}_K : \tilde{X}/K \to \tilde{X}/K$ is called a <u>lifting</u> of $f : X \to X$ if $p_K \circ \tilde{f}_K = f_K \circ p_K$. Two liftings $\tilde{f}_K$ and $\tilde{f}'_K$ are <u>conjugate</u> if there is a $\gamma_K \in \mathcal{D}_K = \pi_1(X)/K$ such that $\tilde{f}'_K = \gamma_K \circ \tilde{f}_K \circ \gamma_K^{-1}$. The conjugacy class of $\tilde{f}_K$, denoted $[\tilde{f}_K]$, is called a <u>lifting class on</u> $\tilde{X}/K$. The subset $p_K \mathrm{Fix}(\tilde{f}_K) \subset \mathrm{Fix}(f)$ is called the <u>mod K fixed point class</u> of f labeled by the lifting class $[\tilde{f}_K]$ on $\tilde{X}/K$. The number of lifting classes on $\tilde{X}/K$ (i.e. the number of mod K fixed point classes) is called the <u>mod K Reidemeister number</u>, written $R_K(f)$.

2.2 THEOREM. The fixed point set $\mathrm{Fix}(f)$ splits into a disjoint union of mod K fixed point classes. Two fixed points x_0 and x_1 belong to the same mod K class iff there is a path c from x_0 to x_1 such that $c(f \circ c)^{-1} \in K$. Each mod K fixed point class is a disjoint union of ordinary fixed point classes. So the index of a mod K fixed point class can be defined in the obvious way. □

2.3 DEFINITION. A mod K fixed point class is <u>essential</u> if its index is nonzero. The number of essential mod K classes is called the <u>mod K Nielsen number</u> of f, denoted $N_K(f)$.

2.4 DEFINITION. Let $\{f_t\}_{t \in I} : X \to X$ be a homotopy. If $\{f_t\}_{t \in I}$ lifts to a homotopy $\{\tilde{f}_{tK}\} : \tilde{X}/K \to \tilde{X}/K$, we say $[\tilde{f}_{0K}]$ corresponds to $[\tilde{f}_{1K}]$ via $\{f_t\}_{t \in I}$, and the mod K fixed point class $p_K \mathrm{Fix}(\tilde{f}_{0K})$ corresponds to $p_K \mathrm{Fix}(\tilde{f}_{1K})$ via $\{f_t\}_{t \in I}$.

2.5 THEOREM. Every homotopy $\{f_t\}_{t \in I} : X \to X$ determines a 1-1 correspondence from lifting classes on $\tilde{X}/K$ (and mod K fixed point classes) of f_0 to those of f_1. This correspondence is index-preserving. Hence $N_K(f)$ is a homotopy invariant. □

2.6 THEOREM. Suppose $f: X \to Y$ and $g: Y \to X$, with $K \lhd \pi_1(X)$ and $K' \lhd \pi_1(Y)$ such that $f_\pi(K) \subset K'$ and $g_\pi(K') \subset K$. Then the correspondence $[\tilde{g}_{K'} \circ \tilde{f}_K] \mapsto [\tilde{f}_K \circ \tilde{g}_{K'}]$ from lifting classes of $g \circ f$ on $\tilde{X}/K$ to lifting classes of $f \circ g$ on $\tilde{Y}/K'$ is 1-1 and index preserving. Hence $N_K(g \circ f) = N_K(f \circ g)$. □

There is another view of the mod K notions which is often more convenient. Look at the covering triangle

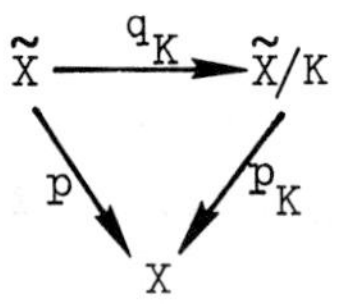

again. Every covering translation $\alpha_K \in \mathcal{D}_K = \pi_1(X)/K$ is the quotient of a covering translation $\alpha \in \mathcal{D} = \pi_1(X)$, by taking equivalence modulo K. And every lifting $\tilde{f}_K : \tilde{X}/K \to \tilde{X}/K$ is induced by (or is the quotient of, or can be lifted to) a lifting $\tilde{f}: \tilde{X} \to \tilde{X}$. Thus, everything on $\tilde{X}/K$ can be considered as induced by something on $\tilde{X}$. Now, we may define two liftings $\tilde{f}, \tilde{f}' : \tilde{X} \to \tilde{X}$ of f to be <u>conjugate mod K</u> if there exist $\gamma \in \pi_1(X)$ and $\kappa \in K$ such that $\tilde{f}' = \kappa \circ \gamma \circ \tilde{f} \circ \gamma^{-1}$, and then define the mod K fixed point class labeled by a mod K conjugacy class of liftings on $\tilde{X}$. This alternative point of view helps to clarify the relationship between ordinary and mod K fixed point classes.

2.7 REMARK. The lifting square

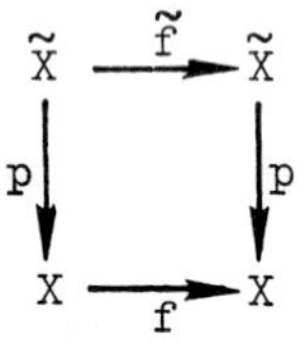

on which the theory of fixed point classes is based splits into two squares.

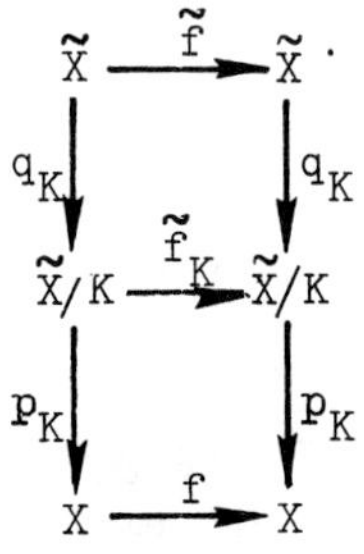

So, in principle, every result about ordinary fixed point classes of a map generalizes in two directions. We may obtain a similar result about the mod K fixed point classes of a map, by routinely taking everything modulo K. And we may consider the ordinary fixed point classes within a given mod K fixed point class, by carefully restricting to a given mod K coset. The results 2.8-11 below may help to illustrate this general principle.

2.8 THEOREM. Let $f : X \to X$ and $K \lhd \pi_1(X)$ be such that $f_\pi K \subset K$. Then $R_K(f) \geq \#\mathrm{Coker}(1 - f_{1*})/\eta \circ \theta(K)$, where $\eta \circ \theta : \pi_1(X) \to \mathrm{Coker}(1 - f_{1*} : H_1(X) \to H_1(X))$ is as in Theorem II.2.1. If $f_\pi^n[\pi_1(X), \pi_1(X)] \subset K$ for some $n > 0$, then equality holds.

PROOF. Our approach in §§II.1-2 works equally well with $\tilde{X}$ and $\pi_1(X)$ replaced by $\tilde{X}/K$ and $\pi_1(X)/K$. This theorem corresponds to Theorems II.2.1 and II.2.5 there. Note that the abelianization of $\pi_1(X)/K$ is nothing but $H_1(X)/\theta(K)$, and

$$\mathrm{Coker}(1 - f_* : H_1(X)/\theta(K) \to H_1(X)/\theta(K)) \cong \mathrm{Coker}(1 - f_{1*})/\eta \circ \theta(K) . \qquad \square$$

2.9 THEOREM. Let $f : X \to X$ and $K \lhd \pi_1(X)$ be such that $f_\pi K \subset K$. Then:

(i) The number of ordinary fixed point classes in any given mod K fixed point class $\geq \#\eta \circ \theta(K)$. If f is eventually commutative, then equality holds.

$(ii)_n$ Suppose $f_\pi^n K$ is in the center of $f_\pi^n \pi_1(X)$ for some $n \geq 0$ (here $f^0 = \mathrm{id}$). If $x_0 \in \mathrm{Fix}(f)$, then the number of ordinary fixed point classes in the mod K fixed point class of x_0 equals $\#(f_\pi^n K/f_\pi^n K \cap f_\pi^n[1])$, where $[1] = \{\gamma f_\pi(\gamma^{-1}) \mid \gamma \in \pi_1(X,x_0)\}$ is the f_π-conjugacy class of 1.

PROOF. Let $\tilde{f}_K : \tilde{X}/K \to \tilde{X}/K$ be a lifting of f, and $\tilde{f} : \tilde{X} \to \tilde{X}$ be a lifting of $\tilde{f}_K$. A lifting $\alpha \circ \tilde{f}$ on $\tilde{X}$, where $\alpha \in \pi_1(X)$, induces on $\tilde{X}/K$ a lifting $\alpha_K \circ \tilde{f}_K$, where α_K is the coset $K\alpha \in \pi_1(X)/K$. It is obvious that $\alpha_K \circ \tilde{f}_K$ is conjugate to $\tilde{f}_K$ iff $\alpha \sim \kappa$ for some $\kappa \in K$, where $\sim$ stands for the $\tilde{f}_\pi$-conjugacy relation. Thus, to count ordinary fixed point classes in the mod K fixed point class of $[\tilde{f}_K]$, we only have to count $\tilde{f}_\pi$-conjugacy classes in K.

(i) Obvious by II.2.1 and II.2.5.

$(ii)_n$ By II.1.3, the assumption is independent of the choice of base point. So take x_0 as base point in X, and take $\tilde{x}_0 \in \mathrm{Fix}(\tilde{f})$ above x_0 as base point in $\tilde{X}$, thus identify $\tilde{f}_\pi$ with f_π by II.1.3. It is trivial to check that $f_\pi^n K \cap f_\pi^n[1]$ is a subgroup of the abelian group $f_\pi^n K$. On the other hand, $\kappa' \sim \kappa \Rightarrow \kappa' = \gamma\kappa f_\pi(\gamma^{-1}) \Rightarrow f_\pi^n(\kappa') = f_\pi^n(\gamma) f_\pi^n(\kappa) f_\pi^{n+1}(\gamma^{-1}) = f_\pi^n(\gamma) f_\pi^{n+1}(\gamma^{-1}) f_\pi^n(\kappa) \Rightarrow f_\pi^n(\kappa') \sim f_\pi^n(\kappa) \Rightarrow \kappa' \sim \kappa$; the last implication is by II.1.6 (ii). Hence $\kappa' \sim \kappa$ iff $f_\pi^n(\kappa'\kappa^{-1}) \in f_\pi^n K \cap f_\pi^n[1]$. $\square$

This theorem describes the loss from counting mod K classes instead of ordinary classes. Note that the count need not be the same for different mod K classes, since the set [1] varies with $\tilde{f}_{\pi}$.

2.10 THEOREM. Suppose $f_{\pi}\pi_1(X) \subset K \cdot J(f)$. Then any two mod K fixed point classes of f have the same index, $R_K(f) = \#\text{Coker}(1 - f_{1*})/\eta \circ \theta(K)$, and $R_K(f)$ divides $R(f)$, $L(f)$ and $N(f)$.

PROOF. Argue as we did for Theorem II.4.1, quoting Theorems 2.5 and 2.8 above instead of II.2.5. □

2.11 THEOREM. Suppose $f_{\pi}K \subset J(f)$. Then any two ordinary fixed point classes in a given mod K fixed point class have the same index, and the counting of 2.9 $(ii)_1$ applies. In particular, if $K \subset J(X)$, then the counting of 2.9 $(ii)_0$ applies.

PROOF. By II.1.3 and II.3.6, $f_{\pi}K \subset J(f)$ implies $\tilde{f}_{\pi}K \subset J(\tilde{f})$ for any lifting $\tilde{f} : \tilde{X} \to \tilde{X}$ of f. As shown in the proof of 2.9, if the ordinary fixed point class of $[\alpha \circ \tilde{f}]$ is in the same mod K fixed point class as $[\tilde{f}]$, then $\alpha \sim \kappa \sim \tilde{f}_{\pi}(\kappa)$ for some $\kappa \in K$, hence $\alpha \sim \tau$ for some $\tau \in J(\tilde{f})$. Thus

$$\begin{aligned} \text{index}(f, p\,\text{Fix}(\alpha \circ f)) &= \text{index}(f, p\,\text{Fix}(\tau \circ \tilde{f})) \\ &= \text{index}(f, p\,\text{Fix}(\tilde{f})) \,, \end{aligned}$$

the last equation is by I.4.5.

The remaining conclusions follow from II.3.7 and 2.9. □

The primary application of the mod K fixed point classes is to estimate $N_K(f)$ as a lower bound of $N(f)$, especially in the case that $\pi_1(X)/K$ is finite, so that $\tilde{X}/K$ is compact. The argument in §II.5 also establishes the following:

2.12 THEOREM. Let $f : X \to X$ and $K \lhd \pi_1(X)$ be such that $f_{\pi}K \subset K$ and $\pi_1(X)/K$ is finite. Then:

(i) For each lifting $\tilde{f}_K : \tilde{X}/K \to \tilde{X}/K$ of f, $L(\tilde{f}_K) \neq 0$ iff $\text{index}(f, p_K\text{Fix}(\tilde{f}_K)) \neq 0$.

(ii) $L(f) = \dfrac{1}{[\pi_1(X) : K]} \sum_{\tilde{f}_K} L(\tilde{f}_K)$, the summation is over all liftings $\tilde{f}_K : \tilde{X}/K \to \tilde{X}/K$.

(iii) Suppose $\tilde{X}/K$ and X have the same rational homology, i.e. $H_*(\tilde{X}/K;Q) \cong H_*(X;Q)$. Then, for any map $f : X \to X$, $L(f) = 0$ implies $N_K(f) = 0$; $L(f) \neq 0$ implies $N_K(f) = R_K(f)$. □

For suitably chosen K, a mod K fixed point class may coincide with an ordinary fixed point class, i.e. a mod K fixed point class may contain only one ordinary fixed point class. Conditions on K may be obtained from Theorem 2.9, but we choose to prove one directly.

2.13 LEMMA. If $K \subset \bigcup_n \operatorname{Ker} f_\pi^n$, then a mod K fixed point class coincides with an ordinary one.

PROOF. By II.1.3, the assumption implies $K \subset \bigcup_n \operatorname{Ker} \tilde{f}_\pi^n$ for every lifting $\tilde{f}: \tilde{X} \to \tilde{X}$. The two notions coincide iff for any $\alpha, \alpha' \in \pi_1(X)$ and any $\kappa \in K$, $\alpha' \sim \kappa\alpha$ implies $\alpha' \sim \alpha$. But, for some n with $\tilde{f}_\pi^n(\kappa) = 1$, by II.1.6 (ii) we have

$$\alpha' \sim \kappa\alpha \sim \tilde{f}_\pi^n(\kappa\alpha) = \tilde{f}_\pi^n(\kappa)\tilde{f}_\pi^n(\alpha) = \tilde{f}_\pi^n(\alpha) \sim \alpha . \qquad \square$$

Consequently, if f is <u>eventually finite</u> in the sense that $f_\pi^n \pi_1(X)$ is finite for some n, or equivalently, $\pi_1(X)/\bigcup_n \operatorname{Ker} f_\pi^n$ is finite, then the ordinary fixed point classes can be analyzed by means of a finite regular covering space.

Now let us put mod K fixed point classes into a functorial framework. It is parallel to §1 in an obvious way.

2.14 DEFINITION. Let us consider the following category. An object is a self-map $f: X \to X$ with a prescribed normal subgroup $K \lhd \pi_1(X)$ such that $f_\pi K \subset K$. A <u>morphism</u>, from $f: X \to X$ with K to $g: Y \to Y$ with K', is a map $h: X \to Y$ such that $h \circ f = g \circ h$ and $h_\pi K \subset K'$. This category will be called <u>the category of self-maps with a normal subgroup</u>.

2.15 DEFINITION. Let X be a compact connected polyhedron, $f: X \to X$ be a map, $K \lhd \pi_1(X)$ be such that $f_\pi K \subset K$. The <u>fixed-point-class data</u> of the pair (f,K), denoted $\mathrm{FPC}_K(f)$ is the weighted set of conjugacy classes of liftings $\tilde{f}_K: \tilde{X}/K \to \tilde{X}/K$ (or, equivalently, of mod K conjugacy classes of liftings $\tilde{f}: \tilde{X} \to \tilde{X}$) of f, the weight of a class $[\tilde{f}_K]$ being the index of the mod K fixed point class $p_K \operatorname{Fix}(\tilde{f}_K)$.

2.16 THEOREM. Suppose $h: X \to Y$ is a morphism in the category of self-maps with a normal subgroup, from $f: X \to X$ with K to $g: Y \to Y$ with K'. Then the correspondence $h_{\mathrm{FPC}}: \mathrm{FPC}(f) \to \mathrm{FPC}(g)$ (defined in Theorem 1.8 (ii)) induces a correspondence $\mathrm{FPC}_K(f) \to \mathrm{FPC}_{K'}(g)$, which we still denote by h_{FPC}. Thus if $h_{\mathrm{FPC}}: [\tilde{f}_K] \mapsto [\tilde{g}_{K'}]$, then $h(p_{X,K}\operatorname{Fix}(\tilde{f}_K)) \subset p_{Y,K'}\operatorname{Fix}(\tilde{g}_{K'})$. The rule that assigns $\mathrm{FPC}_K(f)$ to an object (f,K), and assigns the correspondence $h_{\mathrm{FPC}}: \mathrm{FPC}_K(f) \to \mathrm{FPC}_{K'}(g)$ to a morphism h from (f,K) to (g,K'), is a covariant functor, from the category of self-maps of compact connected polyhedra with a normal subgroup, to the category of weighted sets. This functor will be referred to as the <u>fixed-point-class functor</u>. $\square$

Everything in the preceding section can be generalized to this new fixed-point-class functor. For example, Proposition 1.15 also gives the count of mod $\operatorname{Ker} h_\pi$ fixed point classes, when h_π is not injective, of f which are sent by h into a given fixed point class of g. In other words, it also gives $\# h_{FPC}^{-1}[\tilde{g}]$ for $h_{FPC} : FPC_K(f) \to FPC(g)$ with $K = \operatorname{Ker} h_\pi$. This fact will play an important role in Chapter IV (cf. IV.1.6).

3. FIXED POINT CLASSES OF THE ITERATES. In this section we will discuss the relationship between fixed point classes of f and those of an iterate f^n of f. It aims primarily at a Nielsen type theory for periodic points, which we postpone to the next section. We will end this section with an application to the computation of the Nielsen number $N(f)$ of f itself, improving a result in Chapter II. The Approximation Theorem we use in the proof of 3.6 is of independent interest. It is proved in an appendix.

Let X be a compact connected polyhedron, and let $f : X \to X$ be a map. Let n be a natural number. We will write $\tilde{f}^{(n)}$ for an arbitrary lifting of the iterate $f^n : X \to X$, and write $\tilde{f}^n$ for the iterate of a lifting $\tilde{f} : \tilde{X} \to \tilde{X}$ of f. It is obvious that $p\,\mathrm{Fix}(\tilde{f}) \subset p\,\mathrm{Fix}(\tilde{f}^n)$. So, for a nonempty fixed point class $p\,\mathrm{Fix}(\tilde{f})$ of f, we know $p\,\mathrm{Fix}(\tilde{f}^n)$ is the fixed point class of f^n containing it. If $p\,\mathrm{Fix}(\tilde{f})$ is empty, then, in the set theoretical sense, it is contained in any fixed point class of f^n. But we now want to define $p\,\mathrm{Fix}(\tilde{f}^n)$ to be <u>the</u> fixed point class of f^n containing $p\,\mathrm{Fix}(\tilde{f})$.

3.1 DEFINITION. Let $[\tilde{f}]$ be a lifting class of $f : X \to X$. Then, the lifting class $[\tilde{f}^n]$ of f^n is evidently independent of the choice of representative $\tilde{f}$, so we have a well-defined correspondence

$$\iota : FPC(f) \longrightarrow FPC(f^n) ,$$
$$[\tilde{f}] \longmapsto [\tilde{f}^n] .$$

We say that $p\,\mathrm{Fix}(\tilde{f}^n)$ is <u>the fixed point class of</u> f^n <u>containing</u> $p\,\mathrm{Fix}(\tilde{f})$.

Thus, for $m \mid n$, we also have a correspondence

$$\iota : FPC(f^m) \longrightarrow FPC(f^n) .$$

3.2 PROPOSITION. Let

$$\begin{array}{ccc} X & \xrightarrow{f} & X \\ {\scriptstyle h}\downarrow & & \downarrow{\scriptstyle h} \\ Y & \xrightarrow{g} & Y \end{array} .$$

be a morphism of self-maps. Then, for any n,

$$\begin{array}{ccc} X & \xrightarrow{f^n} & X \\ {\scriptstyle h}\downarrow & & \downarrow{\scriptstyle h} \\ Y & \xrightarrow{g^n} & Y \end{array}$$

is also a morphism of self-maps. And when $m \mid n$, we have a commutative diagram

$$\begin{array}{ccc} FPC(f^m) & \xrightarrow{\iota} & FPC(f^n) \\ {\scriptstyle h_{FPC}}\downarrow & & \downarrow{\scriptstyle h_{FPC}} \\ FPC(g^m) & \xrightarrow{\iota} & FPC(g^n) \end{array}$$

PROOF. Trivial. Recall the definition of h_{FPC} in Theorem 1.8. □

Thus, the fixed point class data of all the iterates of f are organized into a direct system of weighted sets $\{FPC(f^n), \iota\}$, over the set of natural numbers (ordered by divisibility), in the sense of [Eilenberg-Steenrod] Chapter VIII. And, a morphism of self-maps gives rise to a map of direct systems. This generalizes the fixed-point-class functor of 1.10.

As in 1.11, of special importance is the morphism

$$\begin{array}{ccc} X & \xrightarrow{f} & X \\ {\scriptstyle f}\downarrow & & \downarrow{\scriptstyle f} \\ X & \xrightarrow{f} & X \end{array}$$

By Proposition 3.2 above, it induces a correspondence

$$f_{FPC} : FPC(f^n) \to FPC(f^n) .$$

The next proposition gives us an alternative description and shows that it is a built-in automorphism.

3.3 PROPOSITION. (i) Let $\tilde{f}_1, \ldots, \tilde{f}_n$ be liftings of f. Then

$$f_{FPC} : [\tilde{f}_n \circ \cdots \circ \tilde{f}_2 \circ \tilde{f}_1] \mapsto [\tilde{f}_1 \circ \tilde{f}_n \circ \cdots \circ \tilde{f}_2] .$$

(ii) $f(p \operatorname{Fix}(\tilde{f}_n \circ \cdots \circ \tilde{f}_2 \circ \tilde{f}_1)) = p \operatorname{Fix}(\tilde{f}_1 \circ \tilde{f}_n \circ \cdots \circ \tilde{f}_2)$. Thus, the f-image of a fixed point class of f^n is again a fixed point class of f^n, and f_{FPC} indicates the f-image.

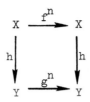

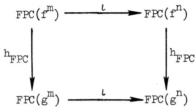

(iii) $\operatorname{index}(f^n, p\operatorname{Fix}(\tilde f_n \circ \cdots \circ \tilde f_2 \circ \tilde f_1)) = \operatorname{index}(f^n, p\operatorname{Fix}(\tilde f_1 \circ \tilde f_n \circ \cdots \circ \tilde f_2))$. In words, f induces an index-preserving permutation among the fixed point classes of f^n.

(iv) $(f_{FPC})^n = \mathrm{id} : FPC(f^n) \to FPC(f^n)$.

PROOF. (i) That every lifting of f^n has a factorization $\tilde f_n \circ \cdots \circ \tilde f_2 \circ \tilde f_1$ follows from Proposition 1.3. Now apply Theorem 1.8 to the commutative diagram of liftings

$$\begin{array}{ccc} \tilde X & \xrightarrow{\tilde f_n \circ \cdots \circ \tilde f_2 \circ \tilde f_1} & \tilde X \\ \Big\downarrow{\scriptstyle \tilde f_1} & & \Big\downarrow{\scriptstyle \tilde f_1} \\ \tilde X & \xrightarrow{\tilde f_1 \circ \tilde f_n \circ \cdots \circ \tilde f_2} & \tilde X \end{array}$$

(ii) Use the obvious equality

$$\tilde f_1 \operatorname{Fix}(\tilde f_n \circ \cdots \circ \tilde f_2 \circ \tilde f_1) = \operatorname{Fix}(\tilde f_1 \circ \tilde f_n \circ \cdots \circ \tilde f_2) .$$

(iii) Apply Theorem 1.12 to

$$X \underset{f^{n-1}}{\overset{f}{\rightleftarrows}} X .$$

(iv) Obvious from (i). □

The direct system $\{FPC(f^n), \iota\}$, equipped with the automorphism f_{FPC}, behaves well with respect to homotopy and commutation.

3.4 THEOREM (Homotopy invariance and commutativity). (i) Suppose $\{f_t\}_{t\in I} : X \to X$ is a homotopy. Then, the correspondence $FPC(f_0^n) \leftrightarrow FPC(f_1^n)$ via the homotopy $\{f_t^n\}$ is an index-preserving bijection, such that the following diagram commutes.

$$\begin{array}{ccccc} FPC(f_0^m) & \xrightarrow{\iota} & FPC(f_0^n) & \xrightarrow{(f_0)_{FPC}} & FPC(f_0^n) \\ \updownarrow & & \updownarrow & & \updownarrow \\ FPC(f_1^m) & \xrightarrow{\iota} & FPC(f_1^n) & \xrightarrow{(f_1)_{FPC}} & FPC(f_1^n) \end{array}$$

(ii) Suppose $f : X \to Y$ and $g : Y \to X$. Then $f_{FPC} : FPC((g \circ f)^n) \to FPC((f \circ g)^n)$ is an index-preserving bijection such that the diagram

$$\begin{array}{ccccc}
FPC((g\circ f)^m) & \xrightarrow{\ \iota\ } & FPC((g\circ f)^n) & \xrightarrow{(g\circ f)_{FPC}} & FPC((g\circ f)^n) \\
\Big\downarrow f_{FPC} & & \Big\downarrow f_{FPC} & & \Big\downarrow f_{FPC} \\
FPC((f\circ g)^m) & \xrightarrow{\ \iota\ } & FPC((f\circ g)^n) & \xrightarrow{(f\circ g)_{FPC}} & FPC((f\circ g)^n)
\end{array}$$

commutes.

PROOF. (i) Recall the morphism

$$\begin{array}{ccc}
X & \xrightarrow{\ f_t\ } & X \\
\Big\downarrow i_t & & \Big\downarrow i_t \\
X\times I & \xrightarrow{\ \mathbb{F}\ } & X\times I
\end{array}$$

introduced before Proposition 1.16. Our correspondence here is nothing but $(i_1)^{-1}_{FPC} \circ (i_0)_{FPC} : FPC(f_0^n) \to FPC(f_1^n)$. So, our theorem follows from 3.2 and the homotopy invariance I.4.5.

(ii) This is a combination of 3.2 and the commutativity 1.12. □

In order to prove the mod p Index Theorem below, we need some preparation. A map $\varphi : X \to Y$ between metric spaces is said to be <u>expanding</u> if $d(\varphi(x),\varphi(x')) > d(x,x')$ whenever $x \neq x'$.

3.5 LEMMA. Let $\varphi : \sigma \to \tau$ be an affine map sending a k-simplex σ onto a k-simplex τ.

(i) If the diameter of σ is less than the distance between the barycenter and the boundary of τ, then φ is expanding.

(ii) If $\sigma \subset \tau$ and φ is expanding, and φ has a fixed point x_0 in the interior of σ, then x_0 is the only fixed point on σ, and

$$\text{index}(\varphi,x_0) = (-1)^k \text{ sgn det } \varphi .$$

PROOF. (i) Let b be the barycenter of σ. For every point $x \in \partial\sigma$, we have $d(b,x) \leq \text{diameter } \sigma < d(\varphi(b),\partial\tau) \leq d(\varphi(b),\varphi(x))$ since $\varphi(x) \in \partial\tau$. Hence φ magnifies distance in every direction.

(ii) An expanding map certainly has at most one fixed point. Take a small sphere S^{k-1} around x_0. Then, on S^{k-1}, the direction field $x \mapsto x - \varphi(x)$ is homotopic to the direction field $x \mapsto x_0 - \varphi(x)$, since $(1 - t)x + tx_0 - \varphi(x)$ never vanishes for $t \in I$ (otherwise, $\varphi(x) = (1-t)x + tx_0$ would imply $\varphi(x) - \varphi(x_0) = (1 - t)x + tx_0 - x_0 = (1 - t)(x - x_0)$, contradicting the assumption that φ is expanding). Hence, by I.3.1,

$$\begin{aligned}\operatorname{index}(\varphi,x_0) &= \text{degree of the direction field } x - \varphi(x)\\ &= \text{degree of the direction field } x_0 - \varphi(x)\\ &= (-1)^k(\text{degree of the direction field } \varphi(x) - \varphi(x_0))\\ &= (-1)^k \operatorname{sgn}\det\varphi .\end{aligned}$$

□

3.6 THEOREM (The mod p Index Theorem). Suppose X is a compact connected polyhedron, and $f: X \to X$ is a map. Suppose $n = p^r$, where p = prime. Let $\mathbb{F}^{(n)}$ be a fixed point class of f^n such that $f\mathbb{F}^{(n)} = \mathbb{F}^{(n)}$. Then

$$\operatorname{index}(f^n,\mathbb{F}^{(n)}) \equiv \sum_i \operatorname{index}(f,\mathbb{F}_i) \mod p ,$$

where the summation is over all fixed point classes $\mathbb{F}_i$ of f contained in $\mathbb{F}^{(n)}$.

PROOF. In view of the Approximation Theorem in the Appendix of this section, and Theorem 3.4 (i), we may assume without loss that f is a simplicial map $K' \to K$ such that

(i) every fixed point of f lies in the interior of some maximal simplex of K' on which f is expanding, and

(ii) f^n has only a finite number of fixed points.

Let $\mathbb{F}_i$ be the fixed point classes of f contained in $\mathbb{F}^{(n)}$. Since $f\mathbb{F}^{(n)} = \mathbb{F}^{(n)}$, then $\mathbb{F}^{(n)} - \cup_i \mathbb{F}_i$ is invariant under f and decomposes into orbits. Each orbit length divides $n = p^r$, hence is divisible by the prime p. But by commutativity, I.3.9, we have

$$\operatorname{index}(f^n,x) = \operatorname{index}(f^n,f(x))$$

for every isolated fixed point x of $f^n = f \circ f^{n-1} = f^{n-1} \circ f$. Hence we have

$$\operatorname{index}(f^n,\mathbb{F}^{(n)}) \equiv \sum_i \operatorname{index}(f^n,\mathbb{F}_i) \mod p .$$

For each $x \in \mathbb{F}_i$, the map f (hence f^n) is an expanding affine map in a k-dimensional Euclidean neighborhood of x. Thus by Lemma 3.5 we have

$$\operatorname{index}(f,x) = (-1)^k \operatorname{sgn}\det f ,$$
$$\operatorname{index}(f^n,x) = (-1)^k \operatorname{sgn}\det f^n .$$

So,

$$\text{index}(f^n,x) = \text{index}(f,x) \qquad \text{if } p \neq 2 ,$$

$$\text{index}(f^n,x) \equiv \text{index}(f,x) \mod 2 \qquad \text{if } p = 2 .$$

In any case

$$\text{index}(f^n,x) \equiv \text{index}(f,x) \mod p .$$

Hence

$$\text{index}(f^n,\mathbb{F}_i) \equiv \text{index}(f,\mathbb{F}_i) \mod p . \qquad \square$$

3.7 COROLLARY. Suppose $n = p^r m$, where p = prime. Let $\mathbb{F}^{(n)}$ be a fixed point class of f^n such that $f^m\mathbb{F}^{(n)} = \mathbb{F}^{(n)}$. Then

$$\text{index}(f^n,\mathbb{F}^{(n)}) \equiv \sum_i \text{index}(f^m,\mathbb{F}_i^{(m)}) \mod p ,$$

where the summation is over all fixed point classes $\mathbb{F}_i^{(m)}$ of f^m contained in $\mathbb{F}^{(n)}$.

PROOF. Apply 3.6 to the map f^m. $\square$

This corollary and Proposition 3.3 (iii) give us conditions on the index (i.e. the weight) that the direct system $\{FPC(f^n),\iota\}$ of fixed-point-class data must satisfy.

In computations, we always pick a lifting $\tilde{f}:\tilde{X}\to\tilde{X}$ of f as reference, and write an arbitrary lifting as $\alpha\circ\tilde{f}$, where $\alpha\in\pi_1(X)$. That is, we use coordinates.

3.8 LEMMA. Let $\tilde{f}:\tilde{X}\to\tilde{X}$ be a lifting of f. Then

$$\iota[\alpha\circ\tilde{f}] = [\alpha^{(n)}\circ\tilde{f}^n]$$

where

$$\alpha^{(n)} = \alpha\tilde{f}_\pi(\alpha)\cdots\tilde{f}_\pi^{n-1}(\alpha) .$$

And

$$f_{FPC}[\beta\circ\tilde{f}^n] = [\tilde{f}_\pi(\beta)\circ\tilde{f}^n] .$$

PROOF. By Definition 3.1, $\iota[\alpha\circ\tilde{f}] = [(\alpha\circ\tilde{f})^n]$. But

$$(\alpha\circ\tilde{f})^n = \alpha\circ\tilde{f}\circ\cdots\circ\alpha\circ\tilde{f}\circ\alpha\circ\tilde{f} = \alpha\circ\tilde{f}\circ\cdots\circ(\alpha\tilde{f}_\pi(\alpha))\circ\tilde{f}\circ\tilde{f} = \cdots$$

$$= (\alpha\tilde{f}_\pi(\alpha)\cdots\tilde{f}_\pi^{n-1}(\alpha))\circ\tilde{f}\circ\cdots\circ\tilde{f}\circ\tilde{f} = \alpha^{(n)}\circ\tilde{f}^n .$$

By 3.3 (i), we have $f_{FPC}[\alpha\circ\tilde{f}^n] = [\tilde{f}\circ(\alpha\circ\tilde{f})\circ\cdots\circ\tilde{f}] = [\tilde{f}_\pi(\alpha)\circ\tilde{f}^n]$. $\square$

3.9 REMARK. All that has been said in this section works also for fixed point classes modulo a normal subgroup $K \lhd \pi_1(X)$. Obviously $f_\pi K \subset K$ implies $f_\pi^n K \subset K$ for all n. So if we can talk about mod K fixed point classes of f, then we can talk about mod K fixed point classes of all iterates of f. We then obtain a direct system of weighted sets $\{FPC_K(f^n),\iota\}$. And, if $h: X \to Y$ is a morphism from a pair (f,K) to (g,K') in the sense of 2.14, then we have a map of direct systems $h_{FPC}: \{FPC_K(f^n),\iota\} \to \{FPC_{K'}(g^n),\iota\}$. This is a generalized form of 2.15-16.

We conclude this section with an application to the computation of $N(f)$, improving Theorems II.4.1-2.

3.10 THEOREM. Let X be a compact connected polyhedron, and let $f: X \to X$ be a map. Suppose there is an integer n such that $f_\pi^n \pi_1(X) \subset J(f^n)$. Then any two fixed point classes of f have the same index. Hence $L(f) = 0$ implies $N(f) = 0$, while $L(f) \neq 0$ implies $N(f) = \#\mathrm{Coker}(1 - f_{1*})$.

PROOF. The case $n = 1$ is just Theorem II.4.1-2. We will use it to prove the general case.

Applying Lemma II.3.7 to the map f^n, we see $f_\pi^n \pi_1(X)$ is abelian, so that f is eventually commutative (cf. Definition II.2.4).

It is clear that $f_\pi^m \pi_1(X) \subset J(f^m)$ for $m > n$, by II.3.8. So we can pick a prime q such that

(a) $f_\pi^q \pi_1(X) \subset J(f^q)$,

(b) q is coprime to the order of the torsion subgroup of $\mathrm{Coker}(1 - f_*)$,

and

(c) q is larger than the absolute value of the difference of indices for any two fixed point classes of f.

Let $\mathbb{F}_i = p\,\mathrm{Fix}(\alpha_i \circ \tilde{f})$, $i = 1,2$, to be two fixed point classes of f. According to Lemma 3.8, the respective fixed point classes of f^q containing them are $\mathbb{F}_i^{(q)} = p\,\mathrm{Fix}(\alpha_i^{(q)} \circ \tilde{f}^q)$, where

$$\alpha_i^{(q)} = \alpha_i \tilde{f}_\pi(\alpha_i) \cdots \tilde{f}_\pi^{q-1}(\alpha_i), \qquad i = 1,2 .$$

We claim that different fixed point classes of f are contained in different fixed point classes of f^q, that is, $\mathbb{F}_1^{(q)} = \mathbb{F}_2^{(q)}$ implies $\mathbb{F}_1 = \mathbb{F}_2$. In fact, suppose $\mathbb{F}_1^{(q)} = \mathbb{F}_2^{(q)}$, then $\alpha_1^{(q)}$ and $\alpha_2^{(q)}$ are $\tilde{f}_\pi^q$-conjugate, i.e. there is a $\gamma \in \pi_1(X)$ such that $\alpha_2^{(q)} = \gamma\alpha_1^{(q)}\tilde{f}_\pi^q(\gamma^{-1})$. Thus, under the epimorphism $\eta \circ \theta: \pi_1(X) \to \mathrm{Coker}(1 - f_*)$ of II.2.1, we have

$$\eta \circ \theta(\alpha_2^{(q)}) = \eta \circ \theta(\alpha_1^{(q)}) + \eta(\theta(\gamma) - f_*^q \circ \theta(\gamma)) = \eta \circ \theta(\alpha_1^{(q)}) .$$

But

$$\eta \circ \theta(\alpha_i^{(q)}) = \eta \circ \theta(\alpha_i \tilde{f}_\pi(\alpha_i) \cdots \tilde{f}_\pi^{q-1}(\alpha_i))$$

$$= \eta(\theta(\alpha_i) + f_* \circ \theta(\alpha_i) + \cdots + f_*^{q-1} \circ \theta(\alpha_i))$$

$$= q\eta \circ \theta(\alpha_i) .$$

So $q\eta \circ \theta(\alpha_1) = q\eta \circ \theta(\alpha_2)$. By the condition (b) for q, we have $\eta \circ \theta(\alpha_1) = \eta \circ \theta(\alpha_2)$. Now we see $\mathbb{F}_1 = \mathbb{F}_2$ by II.2.5. The claim is thus proved.

Applying the case $n = 1$ to the map $f^q : X \to X$, in view of the condition (a) for q, we get

$$\mathrm{index}(f^q, \mathbb{F}_1^{(q)}) = \mathrm{index}(f^q, \mathbb{F}_2^{(q)}) .$$

By the claim above and the mod p Index Theorem, ($f\mathbb{F}_i^{(q)} = \mathbb{F}_i^{(q)}$ since $\mathbb{F}_i \subset \mathbb{F}_i^{(q)}$)

$$\mathrm{index}(f^q, \mathbb{F}_i^{(q)}) \equiv \mathrm{index}(f, \mathbb{F}_i) \mod q, \qquad i = 1,2 .$$

Hence

$$\mathrm{index}(f, \mathbb{F}_1) \equiv \mathrm{index}(f, \mathbb{F}_2) \mod q .$$

But q is large enough by condition (c) so that we have

$$\mathrm{index}(f, \mathbb{F}_1) = \mathrm{index}(f, \mathbb{F}_2) .$$

Hence any two fixed point classes have the same index.

The last sentence of the theorem now follows from II.2.5. □

3.11 COROLLARY. Suppose X is aspherical, and $f : X \to X$ is eventually commutative. Then $L(f) = 0$ implies $N(f) = 0$, while $L(f) \neq 0$ implies $N(f) = \#\mathrm{Coker}(1 - f_*)$.

PROOF. If $f_\pi^n \pi_1(X)$ is abelian, then $f_\pi^n \pi_1(X) \subset J(f^n)$ by Theorem II.3.13. Then apply the preceding theorem. □

Note that this corollary applies to compact surfaces of nonpositive Euler characteristic.

The following example shows that Theorem 3.10 is indeed better than Theorem II.4.1-2.

EXAMPLE. Let $X = \bigvee_{i=1}^{m} S_i$ be a bouquet of circles at x_0. Suppose a map $f : X, x_0 \to X, x_0$ satisfies

$$f(\alpha_1) = k\alpha_1, \qquad (k \neq 0)$$

$$f(\alpha_i) = \alpha_{i-1} \qquad \text{for } i > 1 ,$$

where α_i is a generator of $\pi_1(S_i,x_0) \subset \pi_1(X,x_0)$. Since X is aspherical, by Theorem II.3.13 we have

$$J(f^n) = Z(f^n_\pi \pi_1(X),\pi_1(X)) = \begin{cases} 0 & \text{if } n < m-1, \\ \pi_1(S_1,x_0) & \text{if } n \geq m-1. \end{cases}$$

So that $f^n_\pi \pi_1(X) \not\subset J(f^n)$ for $n < m-1$, but $f^n_\pi \pi_1(X) \subset J(f^n)$ for $n \geq m-1$.

A nice feature of 3.10 which was lacking in II.4.1-2 is that the hypothesis is symmetric (or commutative) in the sense that given $f: X \to Y$ and $g: Y \to X$, then $g \circ f: X \to X$ satisfies the condition iff $f \circ g: Y \to Y$ does. In fact, let $\tilde{f}: \tilde{X} \to \tilde{Y}$ and $\tilde{g}: \tilde{Y} \to \tilde{X}$ be liftings of f, g to the universal coverings, we have

$$(\tilde{f} \circ \tilde{g})^{n+1}_\pi \pi_1(Y) = \tilde{f}_\pi \circ (\tilde{g} \circ \tilde{f})^n_\pi \circ \tilde{g}_\pi \pi_1(Y) \subset \tilde{f}_\pi \circ (\tilde{g} \circ \tilde{f})^n_\pi \pi_1(X)$$

and

$$\tilde{f}_\pi J((\tilde{g} \circ \tilde{f})^n) \subset J(\tilde{f} \circ (\tilde{g} \circ \tilde{f})^n) \subset J(\tilde{f} \circ (\tilde{g} \circ \tilde{f})^n \circ \tilde{g}) = J((\tilde{f} \circ \tilde{g})^{n+1})$$

by Lemma 1.18. Hence $(\tilde{g} \circ \tilde{f})^n_\pi \pi_1(X) \subset J((\tilde{g} \circ \tilde{f})^n)$ implies $(\tilde{f} \circ \tilde{g})^{n+1}_\pi \pi_1(Y) \subset J((\tilde{f} \circ \tilde{g})^{n+1})$.

Recall that commutativity is a built-in symmetry of the fixed point problem. It occurs at both the geometric level, because $\text{Fix}(g \circ f)$ and $\text{Fix}(f \circ g)$ are homeomorphic, and at the algebraic level, as $L(g \circ f) = L(f \circ g)$ and $N(g \circ f) = N(f \circ g)$. But $J(g \circ f) \neq J(f \circ g)$ in general. Theorem 3.10 is the result of an attempt to find a symmetric form for theorems involving $J(f)$. A symmetric form of the divisibility result II.4.9 is yet to be found.

APPENDIX TO §3

APPROXIMATION THEOREM. Let K be a finite simplicial complex, $f: |K| \to |K|$ be a map. Then there is a subdivision K' of K and a simplicial map $g: K' \to K$ homotopic to f such that

(i) every fixed point of g lies in the interior of some maximal simplex of K' on which g is expanding, and

(ii) every iterate g^n of g has only a finite number of fixed points.

PROOF. Each simplex of K of positive dimension has a nonzero distance between its barycenter and its boundary. Let ε be the shortest of such distances. The Hopf Approximation Theorem (cf. [Brown (1971)] p. 118) guarantees the existence of a subdivision K' of K with mesh $K' < \varepsilon$ and a simplicial map $g: K' \to K$ homotopic to f such that every fixed point of

g lies in the interior of some maximal simplex of K'. By Lemma 3.5 (i) and the choice of ε, the map g is expanding on those simplexes where it is nondegenerate. So g satisfies (i). It remains to show (ii).

Let $x \in |K|$ be a fixed point of g^n. Let σ_j be the carrier of $g^j(x)$ in K', $j = 0,1,\dots,n$. Since $g(g^j(x)) = g^{j+1}(x)$ for $j < n$ and $g^n(x) = x$, we have $g(\sigma_j) \supset \sigma_{j+1}$ for $j < n$, and $\sigma_n = \sigma_0$. Hence each $g \mid \sigma_j$ does not decrease dimension (otherwise we can never get $\sigma_n = \sigma_0$), that is, $g \mid \sigma_j$ is nondegenerate for $0 \leq j \leq n - 1$. By the choice of ε, $g \mid \sigma_j$ is expanding for $0 \leq j \leq n - 1$.

Let $\tau = (g \mid \sigma_0)^{-1} \circ (g \mid \sigma_1)^{-1} \circ \cdots \circ (g \mid \sigma_{n-1})^{-1}\sigma_0$, then τ is a simplex in σ_0 containing x. Since every $g \mid \sigma_j$ is expanding, $g^n : \tau \to \sigma_0$ is also expanding, hence x is the only fixed point on τ.

This analysis shows that no two fixed points of g^n can share the same sequence $\sigma_0, \sigma_1, \dots, \sigma_{n-1}$ of simplexes of K'. (If they did, they would be in the same τ, a contradiction.) But K' is a finite complex, so there are only finitely many different sequences of simplexes of length n. Hence g^n has only finitely many fixed points. □

4. THE LEAST NUMBER OF PERIODIC POINTS. Periodic points of a self-map $f : X \to X$, i.e. fixed points of iterates of f, are of importance in several branches of mathematics. We propose to develop a Nielsen type theory for them. The notions in the preceding section were developed primarily for this purpose. The central problem is the estimation from below of the number of periodic points.

4.1 DEFINITION. Let X be a compact connected polyhedron. Let $f : X \to X$ be a self-map. Let

$$P_n(f) = \mathrm{Fix}(f^n) - \bigcup_{m<n} \mathrm{Fix}(f^m)$$

denote the set of periodic points of least period n. Define

$$MP_n[f] := \mathrm{Min}\{\# P_n(g) \mid g \simeq f\} ,$$

$$MF_n[f] := \mathrm{Min}\{\# \mathrm{Fix}(g^n) \mid g \simeq f\} .$$

In words, $MP_n[f]$ is the least number of periodic points of <u>least</u> period n, and $MF_n[f]$ is the least number of periodic points of period n (i.e. fixed points of the n-th iterate), both in the homotopy class $[f]$ of f. They are non-negative integers, in view of the Approximation Theorem in the Appendix to §3.

Note that $MP_1[f] = MF_1[f]$, and it coincides with $MF[f]$ defined in §I.6.

It is evident from this definition that $MF_n[f] \geq MF[f^n]$ and $MF_n[f] \geq \sum_{m|n} MP_m[f]$.

Our first group of examples show various phenomena concerning these least numbers.

EXAMPLE 1. Consider $f : S^{2k+1} \to S^{2k+1}$, $k > 0$, of degree d. If $d \neq 1$, then

$$MF_n[f] = 1 \quad \text{for all } n ,$$

$$MP_n[f] = 0 \quad \text{for all } n > 1 .$$

If $d = 1$, then $MF_n[f] = MP_n[f] = 0$ for all n. It is easy to construct a map realizing all the least numbers $MF_n[f]$ and $MP_n[f]$ simultaneously. Note that when $d = -1$, we have $MF_2[f] > MF[f^2]$.

EXAMPLE 2. Consider $f : S^{2k} \to S^{2k}$ of degree d. If $d \neq -1$, then

$$MF_n[f] = 1 \quad \text{for all } n ,$$

$$MP_n[f] = 0 \quad \text{for all } n > 1 .$$

If $d = -1$, then

$$MF_n[f] = \begin{cases} 0, & \text{if } n \text{ is odd} , \\ 1, & \text{if } n \text{ is even} , \end{cases}$$

$$MP_n[f] = 0 \quad \text{for all } n .$$

In this latter case $MF_2[f] > MP_1[f] + MP_2[f]$, and no map $g \simeq f$ can realize the least numbers $MF_1[f]$ and $MF_2[f]$ (or $MP_1[f]$ and $MP_2[f]$) at the same time.

EXAMPLE 3. Consider self-maps of the k-torus T^k, $k \geq 1$, presented as $\mathbb{R}^k$ modulo the lattice of integral points. Let A be a $k \times k$ integral matrix, $\tilde{f} : \mathbb{R}^k \to \mathbb{R}^k$ be the linear map determined by A, and $f : T^k \to T^k$ be the map covered by $\tilde{f}$. If no eigenvalue of A is a root of unity, then

$$MF_n[f] = MF[f^n] = N(f^n) = |\det(E - A^n)|$$

(cf. Example 2 in §II.4). This f realizes all the least numbers $MF_n[f]$ and $MP_n[f]$ at the same time. No simple formula for $MP_n[f]$ is available. Cf. [Halpern (1979)].

For instance, if $n = 2$ and $A = \begin{pmatrix} 2 & 1 \\ 1 & 1 \end{pmatrix}$, then every map homotopic to f must have an infinite number of periodic points since $|\det(E - A^n)|$ is easily seen to be unbounded.

Example 2 above is instructive. It shows the importance of the reducibility of a periodic point class. The essential class of period 2 is reducible into an inessential class of period 1; this is why $MF_2[f] > MP_1[f] + MP_2[f]$. Let us explain our terminology first.

4.2 DEFINITION. A *periodic point class of period* n of $f: X \to X$ is synonymous with a fixed point class of f^n. The direct system $\{FPC(f^n), \iota\}$ of all the fixed-point-class data, along with the built-in automorphism f_{FPC}, will sometimes be referred to as the *periodic-point-class data* of f. A periodic point class of period n is *reducible to period* $m < n$ if it contains (in the sense of 3.1) some periodic point class of period m; it is *irreducible* if it is not reducible to any lower period. The set of periodic point classes decomposes into f-*orbits* under the action of the automorphism f_{FPC}.

It follows easily from Proposition 3.2 that if a periodic point class is reducible to period m, so is its f-image. Thus, the reducibility (or irreducibility) is a property of an f-orbit.

The reducibility analysis is difficult in general, because the algebraic criterion 3.8 is awkward to use. However, there is an easier case.

4.3 LEMMA. Suppose $\pi_1(X)$ is abelian so that we identify it with $H_1(X)$ and use additive notation. Let $\tilde{f}: \tilde{X} \to \tilde{X}$ be a lifting of $f: X \to X$.

(i) For every n, we have a bijection

$$FPC(f^n) \longleftrightarrow \mathrm{Coker}(1 - f_*^n)$$

such that $[\alpha \circ \tilde{f}^n] \in FPC(f^n)$, where $\alpha \in \pi_1(X)$, corresponds to the coset $[\alpha]^{(n)} = \alpha + \mathrm{Im}(1 - f_*^n) \in \mathrm{Coker}(1 - f_*^n)$. We will call $[\alpha]^{(n)}$ the coordinate of the class $[\alpha \circ \tilde{f}^n]$ with respect to the reference lifting $\tilde{f}$.

(ii) For $m \mid n$, we have a commutative diagram

$$\begin{array}{ccc} FPC(f^m) & \xrightarrow{\ \iota\ } & FPC(f^n) \\ \updownarrow & & \updownarrow \\ \mathrm{Coker}(1-f_*^m) & \xrightarrow{\ \zeta\ } & \mathrm{Coker}(1-f_*^n) \end{array},$$

where ζ is the homomorphism induced by $1 + f_*^m + \cdots + f_*^{n-m}: H_1(X) \to H_1(X)$, i.e., $\zeta[\beta]^{(m)} = [(1 + f_*^m + \cdots + f_*^{n-m})\beta]^{(n)}$.

(iii) The periodic point class with coordinate $[\alpha]^{(n)}$ is reducible to period m iff $\alpha = (1 + f_*^m + \cdots + f_*^{n-m})\beta$ for some β.

(iv) The f-image of the class with coordinate $[\alpha]^{(n)}$ is the one with coordinate $[f_*(\alpha)]^{(n)}$.

PROOF. (i) This is nothing but the coordinate introduced in II.1.8. Since $\pi_1(X)$ is now identified with $H_1(X)$, an $\tilde{f}^n_\pi$-conjugacy class in $\pi_1(X)$ is identified with a coset in $H_1(X)$ mod $\mathrm{Im}(1 - f^n_*)$.

(ii) This follows from Lemma 3.8 by changing the notation from multiplication to addition.

(iii) Obvious from (ii).

(iv) This is just the second half of 3.8. □

4.4 PROPOSITION. Suppose $\pi_1(X)$ is abelian. Suppose a periodic point class of period n is reducible to period ℓ as well as to period m. Then it is reducible to period $k = (\ell,m)$, the greatest common divisor.

PROOF. Without loss we may assume $k = (\ell,m) = 1$. Otherwise, we can set $\ell' = \ell/k$ and $m' = m/k$ and apply the coprime case to the map f^k.

Fix the reference lifting $\tilde{f}$. Suppose the class in question has coordinate $[\alpha]^{(n)}$. According to Lemma 4.3 (iii), there are $\beta, \gamma \in \pi_1(X)$ such that

$$\alpha = (1 + f^\ell_* + \cdots + f^{n-\ell}_*)\beta = (1 + f^m_* + \cdots + f^{n-m}_*)\gamma .$$

Since $(\ell,m) = 1$, it can be shown by the Euclidean algorithm that in the polynomial ring $\mathbb{Z}[t]$, there exist polynomials $B(t)$ and $C(t)$ such that

$$B(t) \cdot (1 + t + \cdots + t^{\ell-1}) + C(t) \cdot (1 + t + \cdots + t^{m-1}) = 1 .$$

Set $\delta = B(f_*)\beta + C(f_*)\gamma$. Then

$$\begin{aligned}(1 + f_* + \cdots + f^{n-1}_*)\delta &= B(f_*)(1 + f_* + \cdots + f^{\ell-1}_*)(1 + f^\ell_* + \cdots + f^{n-\ell}_*)\beta \\ &\quad + C(f_*)(1 + f_* + \cdots + f^{m-1}_*)(1 + f^m_* + \cdots + f^{n-m}_*)\gamma \\ &= B(f_*)(1 + f_* + \cdots + f^{\ell-1}_*)\alpha \\ &\quad + C(f_*)(1 + f_* + \cdots + f^{m-1}_*)\alpha \\ &= \alpha .\end{aligned}$$

So, $[\alpha]^{(n)}$ is reducible to period 1 by 4.3 (iii). □

4.5 REMARK. The conclusion of 4.3 and 4.4 remains true if $f: X \to X$ is eventually commutative. Cf. Theorem II.2.5. These results can also be generalized to mod K periodic point classes when $\pi_1(X)/K$ is abelian. That 4.4 does not apply to general nonabelian $\pi_1(X)$ will be shown by Example 5.

There is a geometric criterion which characterizes the reducibility of a nonempty periodic point class.

4.6 LEMMA. Suppose x is a periodic point of period n, i.e. $f^n(x) = x$. Then, the periodic point class of period n which x belongs to is reducible to period 1 iff there exists a path w from x to $f(x)$ such that the loop

$$wf(w)f^2(w)\cdots f^{n-1}(w)$$

is contractible in X.

PROOF. It is evident from Definitions 4.2 and 3.1 that the class in question is reducible iff $x \in p\,\mathrm{Fix}(\tilde{f}^n)$ for some lifting $\tilde{f}: \tilde{X} \to \tilde{X}$ of f to the universal covering space $\tilde{X}$ of X.

Suppose $x \in p\,\mathrm{Fix}(\tilde{f}^n)$. There is $\tilde{x} \in p^{-1}(x)$ with $\tilde{f}^n(\tilde{x}) = \tilde{x}$. Pick a path $\tilde{w}$ in $\tilde{X}$ from $\tilde{x}$ to $\tilde{f}(\tilde{x})$, and set $w = p \circ \tilde{w}$. Then $\tilde{w}\tilde{f}(\tilde{w})\cdots\tilde{f}^{n-1}(\tilde{w})$ is a path from $\tilde{x}$ to $\tilde{f}^n(\tilde{x})$, hence a loop in $\tilde{X}$. Therefore its projection $wf(w)\cdots f^{n-1}(w)$ is homotopically trivial in X.

On the other hand, suppose there is a path w from x to $f(x)$ such that $wf(w)\cdots f^{n-1}(w)$ is a homotopically trivial loop in X. Pick a point $\tilde{x} \in p^{-1}(x)$, lift w to a path $\tilde{w}$ in $\tilde{X}$ starting from $\tilde{x}$ and ending in a point $\tilde{y} \in p^{-1}(f(x))$. Let $\tilde{f}$ be the lifting of f such that $\tilde{f}(\tilde{x}) = \tilde{y}$. Then $\tilde{w}\tilde{f}(\tilde{w})\cdots\tilde{f}^{n-1}(\tilde{w})$ is a path in $\tilde{X}$ from $\tilde{x}$ to $\tilde{f}^n(\tilde{x})$. But it is a lifting of a contractible loop $wf(w)\cdots f^{n-1}(w)$ in X, so it is a loop in $\tilde{X}$, i.e. $\tilde{f}^n(\tilde{x}) = \tilde{x}$. Thus $x \in p\,\mathrm{Fix}(\tilde{f}^n)$. □

The next two examples show the application of the algebraic and geometric criteria of reducibility.

EXAMPLE 4. Let $X = \mathbb{RP}^3$, obtained from S^3 by identifying antipodal points. Thus the projection $p: S^3 \to \mathbb{RP}^3$ is the universal covering of X. A map $f: \mathbb{RP}^3 \to \mathbb{RP}^3$ is defined by specifying a lifting $f: S^3 \to S^3$ as follows.

$$S^3 = \{(r_1e^{i\theta_1}, r_2e^{i\theta_2}) \mid r_1^2 + r_2^2 = 1\},$$

$$\tilde{f}(r_1e^{i\theta_1}, r_2e^{i\theta_2}) = (r_1e^{i3\theta_1}, r_2e^{i3\theta_2}).$$

It is obvious that $\deg f = 9$ and $f_\pi = \mathrm{id}: \mathbb{Z}_2 \to \mathbb{Z}_2$. Hence $L(f^n) = 1 - 9^n < 0$, and $R(f^n) = \#\mathrm{Coker}(1 - f^n_{1*}) = 2$ for all n. By II.4.4, we know $N(f^n) = 2$ for all n.

Let us examine the reducibility of the two essential fixed point classes of f^2. Since $\pi_1(\mathbb{RP}^3) \cong \mathbb{Z}_2$ is abelian and $1 + f_\pi = 0$, it follows from Lemma 4.3 (iii) that only one fixed point class, namely the one labeled by $[\tilde{f}^2]$, is reducible, Denote this one by $\mathbb{F}_0^{(2)}$, the other one by $\mathbb{F}_1^{(2)}$. Denote the fixed point classes of f by $\mathbb{F}_0$ and $\mathbb{F}_1$. Thus $\mathbb{F}_0^{(2)} \supset \mathbb{F}_0 \cup \mathbb{F}_1$, while $\mathbb{F}_1^{(2)}$ is irreducible and $f\mathbb{F}_1^{(2)} = \mathbb{F}_1^{(2)}$.

This analysis holds as well for any map g homotopic to f, because all the notions used have homotopy invariance. So, for such a map g, the iterate g^2 has at least two fixed points in $\mathbb{F}_0^{(2)}$, one in $\mathbb{F}_0$ and one in $\mathbb{F}_1$. And g^2 also has at least two fixed points in the irreducible $\mathbb{F}_1^{(2)}$, since a g-orbit in $\mathbb{F}_1^{(2)}$ would not reduce to a single point. Thus, $MP_2[f] \geq 2$ and $MF_2[f] \geq 4$. By induction it is not hard to see $MF_{2^r}[f] \geq 2^{r+1}$ and $MP_{2^r}[f] \geq 2^r$. They are much larger than $N(f^{2^r}) = 2$.

We may use the same idea to construct mappings on generalized lens spaces with interesting estimates of $MP_n[f]$ and $MF_n[f]$.

EXAMPLE 5. (The figure Θ). Let $A = \{a_1, a_2\}$ and $B = \{b_1, b_2, b_3\}$, and let X be the join $A * B$. Then X is a one dimensional simplicial complex homeomorphic to the Greek letter Θ. Let $f : X \to X$ be the simplicial automorphism of X which interchanges a_1, a_2 and cyclically permutes b_1, b_2 and b_3. It is obvious that $\mathrm{Fix}(f) = \emptyset$, $\mathrm{Fix}(f^2) = A$, $\mathrm{Fix}(f^3) = B$, and $\mathrm{Fix}(f^6) = X$. The two points of period 2 are in two essential classes, both of index 1. Also, the three points of period 3 are in three essential classes of index 1. There is only one essential class of period 6, with index $\chi(X) = -1$.

We can easily find a map g homotopic to f, by disturbing f a little bit on one of the edges, such that

$$\mathrm{Fix}(g^n) = \begin{cases} \emptyset, & \text{if } (6,n) = 1 , \\ A, & \text{if } (6,n) = 2 , \\ B, & \text{if } (6,n) = 3 , \\ A \cup B, & \text{if } 6 \mid n . \end{cases}$$

An easy argument shows that g realizes all the least numbers $MF_n[f]$ and $MP_n[f]$.

The point we want to make is that, the essential class of period 6, though reducible to periods 2 and 3, is not reducible to period 1. This contrasts with Proposition 4.4. Proof by contradiction: Suppose, by 4.6, there were an edge path w from a_1 to a_2 such that $wf(w)\cdots f^5(w) \simeq 0$. Denote the three simple arcs from a_1 to a_2 by three letters t, u, v respectively. Then w is a word of odd length, so the length of the word $wf(w)\cdots f^5(w)$ is congruent to 6 mod 12. On the other hand, since $wf(w)\cdots f^5(w) \simeq 0$ and X is one dimensional, this word should be cyclically reducible to the empty word. But in view of the $\mathbb{Z}_6$-symmetry of this word, cyclic cancellation can be done symmetrically, reducing the word length by twelve at a time. So the word length should be congruent to 0 mod 12. A contradiction.

Another observation: In Example 5, the two irreducible essential classes of period 2 give us one period 2 point each, while the single such in Example 4 contributes two period 2 points. Explanation: The two in Example 5 constitute an f-orbit, but the one in Example 4 is an f-orbit by itself.

4.7 DEFINITION. A set of periodic point classes (of diverse periods) is f-invariant if it contains the f-images of its elements, i.e. if it is a union of f-orbits. The height of an f-invariant set of periodic point classes is the sum of the periods of the f-orbits which it decomposes into.

4.8 DEFINITION. The Nielsen type number of period n, denoted $NP_n(f)$, of a self-map $f: X \to X$ is the height of the set of irreducible essential periodic point classes of period n. The Nielsen type number for the n-th iterate, denoted $NF_n(f)$, is the minimal height of such f-invariant sets of periodic point classes, that each essential class of any period $m \mid n$ contains at least one class in the set. These numbers will generally be referred to as the Nielsen type numbers for periodic points.

A procedure for finding $NF_n(f)$ from the periodic-point-class data of f: Take the f-invariant set S of all the essential classes, of any period $m \mid n$, which do not contain any essential classes of lower period. To each f-orbit in S, find the lowest period which it can be reduced to. The sum of these numbers is $NF_n(f)$.

The following inequalities follow directly from the definitions. The reader is invited to write down necessary and sufficient conditions for the equalities to hold.

4.9 PROPOSITION. (i) $NF_n(f) \geq NF_m(f)$ for $m \mid n$.

(ii) $NF_n(f) \geq NF_m(f^{n/m})$ for $m \mid n$.

(iii) $NF_n(f) \geq \sum_{m|n} NP_m(f)$. □

4.10 THEOREM (Homotopy invariance and commutativity).

(i) If $f \simeq g: X \to X$, then $NP_n(f) = NP_n(g)$ and $NF_n(f) = NF_n(g)$ for all n.

(ii) If $f: X \to Y$ and $g: Y \to X$, then $NP_n(g \circ f) = NP_n(f \circ g)$ and $NF_n(g \circ f) = NF_n(f \circ g)$ for all n.

PROOF. The invariants NP_n and NF_n are determined by the periodic-point-class data, so we may appeal to Theorem 3.4. □

4.11 THEOREM. Suppose two self-maps $f: X \to X$ and $g: Y \to Y$ of compact connected polyhedra are of the same homotopy type. Then $NP_n(f) = NP_n(g)$ and $NF_n(f) = NF_n(g)$ for every n.

PROOF. Similar to the proof of Theorem I.5.4. Use the preceding theorem. □

4.12 THEOREM. $NP_n(f) \leq MP_n[f]$ and $NF_n(f) \leq MF_n[f]$. Moreover,

$$NP_n(f) \leq \operatorname{Min}\{\# P_n(g) \mid g \text{ has the same homotopy type as } f\},$$

$$NF_n(f) \leq \operatorname{Min}\{\# \operatorname{Fix}(g^n) \mid g \text{ has the same homotopy type as } f\}.$$

PROOF. In view of the preceding theorem, it suffices to prove $\# P_n(f) \geq NP_n(f)$ and $\# \mathrm{Fix}(f^n) \geq NF_n(f)$. Always keep in mind that an essential class is geometrically non-empty.

Obviously, each f-orbit in $P_n(f)$ consists of n points, and belongs to an f-orbit of periodic point classes of order n. So, $\# P_n(f)$ equals n times the number of f-orbits in $P_n(f)$, hence is greater than or equal to n times the number of essential irreducible f-orbits of period n, i.e. $\# P_n(f) \geq NP_n(f)$.

Every point $x \in \mathrm{Fix}(f^n)$ has a least period $m_x \mid n$, and belongs to a unique periodic point class of period m_x. The totality of the classes thus obtained forms a set S of periodic point classes of f. It is easily seen that S is f-invariant, and every essential class of any period $m \mid n$ must contain at least one class from S, so that the height of S is less than or equal to $NF_n(f)$ by Definition 4.8. On the other hand, the f-orbit length of $x \in \mathrm{Fix}(f^n)$ equals the period, namely m_x, of the containing f-orbit in S. So that $\# \mathrm{Fix}(f^n) \geq$ the height of S. Thus $\# \mathrm{Fix}(f^n) \geq NF_n(f)$. □

How can we compute the Nielsen type invariants $NP_n(f)$ and $NF_n(f)$? Certainly it is no easier than computing the Nielsen number $N(f)$. But in the cases where the computation of $N(f)$ is feasible, such as those studied in Chapter II, we have a fair chance. Moreover, we may, as we did in §2, introduce the notion of mod K periodic point classes and define $NP_{n,K}(f)$ and $NF_{n,K}(f)$ as in Definitions 4.7-8. They are easily seen to be lower bounds for $NP_n(f)$ and $NF_n(f)$, and are sometimes easier to estimate. As an example of this approach, we give the following generalization of Example 4 above.

4.13 THEOREM. Suppose X satisfies one of the following conditions:

(a) $J(X) = \pi_1(X)$, or

(b) $\pi_1(X)$ is finite and $H_*(\tilde{X};Q) \cong H_*(X;Q)$.

Let $f: X \to X$ be a map, and let $k = \# \mathrm{Coker}(1 - f_*)$.

(i) If $L(f^n) \neq 0$, where n has no prime factor other than those of k, then $NP_n(f) \geq k\varphi(n)$, where $\varphi(n)$ is the Euler function in number theory.

(ii) If $L(f^m) \neq 0$ for all $m \mid n$, then $NF_n(f) \geq kn'$, where n' is obtained from n by removing all the prime factors that are absent in k.

(iii). If $k > 1$ and $L(f^n) \neq 0$ for all n, then every map g homotopic to f has an infinite number of periodic points.

PROOF. For brevity let us write $C = \mathrm{Coker}(1 - f_*)$ and $K = \mathrm{Ker}(\eta \circ \theta : \pi_1(X) \to C)$, where $\eta \circ \theta$ is as in II.2.1. The homomorphism $C \to C$ induced by f_π is nothing but the identity. Consider periodic point classes modulo K. By a trivial generalization of Lemma 4.3, we have, for $m \mid n$, a commutative diagram

$$\begin{array}{ccc} FPC_K(f^m) & \xrightarrow{\iota} & FPC_K(f^n) \\ \updownarrow & & \updownarrow \\ C & \xrightarrow{\zeta} & C \end{array}$$

where ζ is multiplication by n/m. And each f-orbit is a single element.

On the other hand, the space X has the nice property that for any map $g : X \to X$, the index of every fixed point class (ordinary or modulo any normal subgroup) has the same sign as $L(g)$. This follows from II.4.4 for (a), and from Remark II.5.7 and the proof of II.5.6 for (b). This greatly simplifies the computation of $NP_{n,K}(f)$ and $NF_{n,K}(f)$, in that we can now count the classes without worrying about their indices.

(i) According to the diagram above, $NP_{n,K}(f)$ equals n times the number of elements of C which are not divisible by any prime factor of n. Let $p_1,\ldots,p_r$ be the prime factors of k, and let $C = C_{p_1} \oplus \cdots \oplus C_{p_r}$ be the decomposition of C into primary components. Suppose $p_1,\ldots,p_s$ are the prime factors of n. Then the set C' of elements indivisible by $p_1,\ldots,p_s$ is $C' = C - \cup_{i=1}^{s} p_i C = (C_{p_1} - p_1 C_{p_1}) \oplus \cdots \oplus (C_{p_s} - p_s C_{p_s}) \oplus \cdots \oplus C_{p_r}$. But $\#(C_{p_i} - p_i C_{p_i}) \geq (1 - \frac{1}{p_i})(\# C_{p_i})$, so that $\# C' \geq (1 - \frac{1}{p_1})\cdots(1 - \frac{1}{p_s})k$. Hence $NP_{n,K}(f) \geq n \cdot k \cdot (1 - \frac{1}{p_1})\cdots(1 - \frac{1}{p_s}) = k\varphi(n)$.

(ii)
$$NF_{n,K}(f) \geq NF_{n',K}(f) \geq \sum_{m \mid n'} NP_{m,K}(f) \qquad \text{(by 4.9)}$$
$$\geq \sum_{m \mid n'} k\varphi(m) = kn' \qquad \text{(by (i))} .$$

(iii) By Theorem 4.12 and (ii) above, we have

$$\# \mathrm{Fix}(g^{k^r}) \geq NF_{k^r,K}(f) \geq k \cdot k^r = k^{r+1} ,$$

unbounded when $r \to \infty$. □

A natural question is whether the lower bounds $NP_n(f)$ and $NF_n(f)$ for $MP_n[f]$ and $MF_n[f]$ can be further improved. The answer is no in general, as B. Halpern has announced the following important results.

4.14 THEOREM. If X is a compact connected differentiable manifold of dimension ≥ 5, then $MP_n[f] = NP_n(f)$ and $MF_n[f] = NF_n(f)$ for all n.

4.15 THEOREM. For all n,

$$NP_n(f) = \mathrm{Min}\{\# P_n(g) \mid g \text{ has the same homotopy type as } f\},$$

$$NF_n(f) = \mathrm{Min}\{\# \mathrm{Fix}(g^n) \mid g \text{ has the same homotopy type as } f\}.$$

Here we restrict the spaces involved to be compact connected ANRs.

See [Halpern (1980)] for the results about $NP_n(f)$. The results about $NF_n(f)$ can be proved by the same method (private communication).

Now we can generalize to periodic points the converses, given in II.6, of the Lefschetz fixed point theorem.

4.16 THEOREM. Suppose X is a compact connected differentiable manifold of dimension ≥ 5, such that either (a) $J(X) = \pi_1(X)$, or (b) $\pi_1(X)$ is finite and $H_*(\tilde{X};\mathbb{Q}) \cong H_*(X;\mathbb{Q})$. Let $f: X \to X$ be a map, and let n be a natural number.

(i) If $L(f^n) = 0$, then there is a map $g \simeq f$ such that g has no periodic point of least period n.

(ii) If $L(f^m) = 0$ for all $m \mid n$, then there is a map $g \simeq f$ such that g^n has no fixed point.

PROOF. It follows from II.4.4 (for (a)) or II.5.6-7 (for (b)) and Definition 4.8 that $NP_n(f) = 0$ or $NF_n(f) = 0$ under the respective assumptions. Then apply Theorem 4.14. □

CHAPTER IV

FIXED POINT CLASSES OF A FIBER MAP

This chapter is an exposition of the Nielsen theory for fiber maps, i.e. maps of fiber spaces which send fibers into fibers. The aim is to study the relation between fixed point classes in the fiber, the base and the total space. Special attention has been paid to product formulas for the Nielsen numbers which are helpful in computations.

Section 1 provides general information on fiber maps, and specializes the results of §§III.1-2 to the present situation. Section 2 is crucial to the theory. It exhibits the fine structure of the intersection of a fiber with a fixed point class in the total space. Section 3 discusses the product formula for the fixed point index in the total space, thus clears the way for the product formulas in §4.

1. FIBER MAPS.

A fiber map is a morphism of self-maps which is a fibration. In this section we first recall some basic facts about fibrations, then put together some observations which are immediate consequences of the results in Chapter III.

A map $q : E \to B$ is a <u>fibration</u> if it has the homotopy lifting property, i.e., for any commutative square

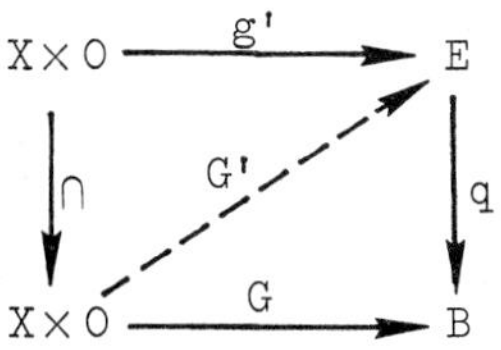

there exists a map represented by the dotted arrow that makes the two triangles commutative. (A standard reference is [Spanier] Chapter 2.) For a fibration $q : E \to B$ there exists a <u>path lifting function</u>, i.e. a continuous map

$$\lambda : W = \{(e,w) \in E \times B^I \mid q(e) = w(0)\} \to E^I$$

such that $\lambda(e,w)(0) = e$ and $q(\lambda(e,w)(t)) = w(t)$ for all $t \in I$. In words, $\lambda(e,w)$ is a lifting of the path w starting from e. For a metric base space B, we may assume without loss of generality that the path lifting function is regular, in the sense that the lifting of a constant path is always a constant path. (Cf. [Hurewicz].)

The subspace $q^{-1}(b)$ of E is called the fiber at $b \in B$, written $F_b = q^{-1}(b)$. A path lifting function gives rise to a <u>fiber translation function</u> $\tau : W = \{(e,w) \in E \times B^I \mid q(e) = w(0)\} \to E$ by $\tau(e,w) = \lambda(e,w)(1)$. Then $\tau(e,w) \in F_{w(1)}$, so $\tau_w = \tau(\cdot, w) : F_{w(0)} \to F_{w(1)}$. Thus a regular path lifting function λ gives rise to a fiber translation function $\tau_w : F_{w(0)} \to F_{w(1)}$ which depends continuously on $w \in B^I$, and $\tau_w = \mathrm{id}$ for a constant path w.

EXERCISE 1. Let $q : E \to B$ be a fibration with metric base B. Prove that $\tau_w : F_{w(0)} \to F_{w(1)}$ is a homotopy equivalence.

EXERCISE 2. Let w and w' be two paths in B such that the product path ww' is defined. Then $\tau_{w'} \circ \tau_w \simeq \tau_{ww'}$.

Let $q : E \to B$ be a fibration with E, B and all fibers path-connected, locally path-connected and semi-locally 1-connected. Let $p : \tilde{E} \to E$ and $\bar{p} : \tilde{B} \to B$ be the universal covering spaces of E and B respectively. Let $\tilde{q} : \tilde{E} \to \tilde{B}$ be a lifting of q. Thus we have a commutative square

$$\begin{array}{ccc} \tilde{E} & \xrightarrow{p} & E \\ \Big\downarrow{\scriptstyle \tilde{q}} & & \Big\downarrow{\scriptstyle q} \\ \tilde{B} & \xrightarrow{\bar{p}} & B \end{array}$$

Let us write $F_b = q^{-1}(b)$, $\tilde{F}_{\tilde{b}} = \tilde{q}^{-1}(\tilde{b})$.

1.1 LEMMA. (i) The lifting $\tilde{q} : \tilde{E} \to \tilde{B}$ is a fibration with path-connected fibers.

(ii) For any $b \in B$, $p^{-1}(F_b) = \coprod_{\tilde{b} \in \bar{p}^{-1}(b)} \tilde{F}_{\tilde{b}}$, i.e. these $\tilde{F}_{\tilde{b}}$ form a path-component decomposition of $p^{-1}(F_b)$. The restriction $p \mid \tilde{F}_{\tilde{b}} : \tilde{F}_{\tilde{b}} \to F_b$ is a regular covering, with $p_\pi \pi_1(\tilde{F}_{\tilde{b}}) = K_b := \mathrm{Ker}(i_\pi : \pi_1(F_b) \to \pi_1(E))$.

(iii) A fiber translation function $\tau_w : F_{w(0)} \to F_{w(1)}$ gives rise to a fiber translation function $\tilde{\tau}_{\tilde{w}} : \tilde{F}_{\tilde{w}(0)} \to \tilde{F}_{\tilde{w}(1)}$, such that $\tilde{\tau}_{\tilde{w}}$ is a lifting of $\tau_{\bar{p} \circ \tilde{w}}$. That is, the diagram

$$\begin{array}{ccc} \tilde{F}_{\tilde{w}(0)} & \xrightarrow{\tilde{\tau}_w} & \tilde{F}_{\tilde{w}(1)} \\ \Big\downarrow{\scriptstyle p} & & \Big\downarrow{\scriptstyle p} \\ F_{w(0)} & \xrightarrow{\tau_w} & F_{w(1)} \end{array}$$

commutes, where $w = \bar{p} \circ \tilde{w}$ is the projection of $\tilde{w}$.

PROOF. (i) We prove $\tilde{q}$ has the homotopy lifting property. Suppose we have maps g' and G making the diagram

$$\begin{array}{ccccc} X \times 0 & \xrightarrow{g'} & \tilde{E} & \xrightarrow{p} & E \\ {\scriptstyle i}\downarrow\cap & \overset{G'}{\nearrow} & \downarrow{\scriptstyle \tilde{q}} & & \downarrow{\scriptstyle q} \\ X \times I & \xrightarrow{G} & \tilde{B} & \xrightarrow{\bar{p}} & B \end{array}$$

commutative. Since p and q are fibrations, so is their composition $q \circ p = \bar{p} \circ \tilde{q}$. So there is a G' such that the upper triangle is commutative, and that $(\bar{p} \circ \tilde{q}) \circ G' = \bar{p} \circ G$. Now both G and $\tilde{q} \circ G'$ are liftings of the same map, and they agree on $X \times 0$, so by the unique lifting property of the covering space $\bar{p} : \tilde{B} \to B$, we have $G = \tilde{q} \circ G'$ as required. The space $\tilde{F}_{\tilde{b}}$ is path-connected since in the homotopy exact sequence

$$\cdots \to \pi_1(\tilde{B}) \to \pi_0(\tilde{F}_{\tilde{b}}) \to \pi_0(\tilde{E}) \to \cdots$$

we already know $\pi_1(\tilde{B}) = \pi_0(\tilde{E}) = 0$.

(ii) Obviously $p^{-1}(F_b) = p^{-1}(q^{-1}(b)) = \tilde{q}^{-1}(\bar{p}^{-1}(b)) = \coprod_{\tilde{b} \in \bar{p}^{-1}(b)} \tilde{F}_{\tilde{b}}$. The restriction of p to each component of $p^{-1}(F_b)$ is certainly a covering. The information about π_1 follows from the following commutative diagram with exact rows.

$$\begin{array}{ccccc} \pi_2(\tilde{B}) & \longrightarrow & \pi_1(\tilde{F}_{\tilde{b}}) & \longrightarrow & \pi_1(\tilde{E}) = 0 \\ \cong\downarrow{\scriptstyle \bar{p}_\pi} & & \downarrow{\scriptstyle p_\pi} & & \downarrow{\scriptstyle p_\pi} \\ \pi_2(B) & \longrightarrow & \pi_1(F_b) & \longrightarrow & \pi_1(E) \end{array}$$

(iii) From the proof of (i) we know a path lifting function

$$\tilde{\lambda} : \tilde{W} = \{(\tilde{e},\tilde{w}) \in \tilde{E} \times \tilde{B}^I \mid \tilde{q}(\tilde{e}) = \tilde{w}(0)\} \to \tilde{E}^I$$

for $\tilde{q} : \tilde{E} \to \tilde{B}$ can be obtained from a path lifting function $\lambda : W \to E^I$ for $q : E \to B$ as follows: $\tilde{\lambda}(\tilde{e},\tilde{w})$ is the lifting in $\tilde{E}$, starting from $\tilde{e}$, of the path $\lambda(p(\tilde{e}), \bar{p} \circ \tilde{w})$ in E. The fiber translation function $\tilde{\tau}$ determined by this $\tilde{\lambda}$ is obviously a lifting of the fiber translation function τ determined by λ. □

1.2 DEFINITION. Let $q : E \to B$ be a fibration. A map $f : E \to E$ is called a <u>fiber map</u> if there is an <u>induced map</u> $\bar{f} : B \to B$ such that

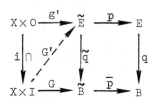

$$\begin{array}{ccc} E & \xrightarrow{f} & E \\ {\scriptstyle q}\downarrow & & \downarrow{\scriptstyle q} \\ B & \xrightarrow{\bar{f}} & B \end{array}$$

is commutative. In other words, a fiber map is a morphism of self-maps which is also a fibration.

As a direct consequence of III.1.5, III.1.8 and III.1.14 (i), we have

1.3 LEMMA. Let $q : E \to B$ be a fibration and $\tilde{q} : \tilde{E} \to \tilde{B}$ be a lifting. Let $f : E \to E$ be a fiber map inducing $\bar{f} : B \to B$. Then every lifting $\tilde{f} : \tilde{E} \to \tilde{E}$ of f is a fiber map inducing a lifting $\tilde{\bar{f}} : \tilde{B} \to \tilde{B}$ of $\bar{f}$. Every fixed point class of f is mapped by q into a fixed point class of $\bar{f}$, namely $q_{FPC}[\tilde{f}] = [\tilde{\bar{f}}]$ and $q(p\ \mathrm{Fix}(\tilde{f})) \subset \bar{p}\ \mathrm{Fix}(\tilde{\bar{f}})$. Furthermore, for every fixed point class of $\bar{f}$, there exists a fixed point class of f that is mapped by q into it. □

Let

$$\begin{array}{ccc} E & \xrightarrow{f} & E \\ {\scriptstyle q}\downarrow & & \downarrow{\scriptstyle q} \\ B & \xrightarrow{\bar{f}} & B \end{array}$$

be a fiber map. The induced map $\bar{f} : B \to B$ is a book-keeping map in the sense that it keeps track of which fiber gets mapped into which fiber, that is, $f(F_b) \subset F_{\bar{f}(b)}$. Suppose $b \in B$ is a fixed point of $\bar{f}$, then $f(F_b) \subset F_b$, so that $f_b = f \mid F_b$ is also a self-map, and the inclusion $i : F_b \to E$ is a morphism of self-maps.

$$\begin{array}{ccc} F_b & \xrightarrow{f_b} & F_b \\ {\scriptstyle i}\downarrow\cap & & \downarrow\cap{\scriptstyle i} \\ E & \xrightarrow{f} & E \end{array}$$

A direct consequence of Theorem III.2.16 is the following.

1.4 LEMMA. Every mod K_b fixed point class of f_b is contained in a fixed point class of f. □

The mod K_b fixed point class of f_b is an important object to study for the following geometric reason. Recall from Lemma 1.1 (ii) that each component

$\tilde{F}_{\tilde{b}}$ of $p^{-1}(F_b)$ is a covering space of F_b with $\pi_1 = K_b$. Let $\tilde{f}: \tilde{E} \to \tilde{E}$ be a lifting of a fiber map $f: E \to E$. Since the induced map $\tilde{\tilde{f}}: \tilde{B} \to \tilde{B}$ keeps track of which fiber $\tilde{F}_{\tilde{b}}$ gets mapped into which fiber, $\tilde{f}(\tilde{F}_{\tilde{b}}) \subset \tilde{F}_{\tilde{\tilde{f}}(\tilde{b})}$, we see $\tilde{f}(\tilde{F}_{\tilde{b}}) \subset \tilde{F}_{\tilde{b}}$ iff $\tilde{b} \in \mathrm{Fix}(\tilde{\tilde{f}})$. In this case, the restriction $\tilde{f}_{\tilde{b}} = \tilde{f} \mid \tilde{F}_{\tilde{b}}$ is a lifting of $f_b: F_b \to F_b$, and is sent into $\tilde{f}$ by the inclusion $\tilde{i}: \tilde{F}_{\tilde{b}} \to \tilde{E}$. In other words, the mod K_b

$$\begin{array}{ccc} \tilde{F}_{\tilde{b}} & \xrightarrow{\tilde{f}_{\tilde{b}}} & \tilde{F}_{\tilde{b}} \\ {\scriptstyle \tilde{i}}\downarrow\cap & & {\scriptstyle \tilde{i}}\downarrow\cap \\ \tilde{E} & \xrightarrow{\tilde{f}} & \tilde{E} \end{array}$$

fixed point class $p\,\mathrm{Fix}(\tilde{f}_{\tilde{b}})$ of f_b is contained in the fixed point class $p\,\mathrm{Fix}(\tilde{f})$ of f. The converse is also true:

1.5 LEMMA. Every mod K_b fixed point class of f_b that is mapped by the inclusion $i: F_b \to E$ into the fixed point class $p\,\mathrm{Fix}(\tilde{f})$ of f presents itself as $p\,\mathrm{Fix}(\tilde{f}_{\tilde{b}})$ for some $\tilde{b} \in \mathrm{Fix}(\tilde{\tilde{f}}) \cap \bar{p}^{-1}(b)$. In other words, these $\tilde{f}_{\tilde{b}}$ represent all (possibly with repetition) such mod K_b lifting classes of f_b.

PROOF. What is to be proved is the following. Let F' be a covering space of F_b with $\pi_1 = K_b$, and let $f': F' \to F'$ be a lifting of $f_b: F_b \to F_b$ such that $i_{FPC}[f'] = [\tilde{f}]$. Then there is a $\tilde{b} \in \bar{p}^{-1}(b) \cap \mathrm{Fix}(\tilde{\tilde{f}})$ and an isomorphism $i': F' \cong \tilde{F}_{\tilde{b}}$ of covering spaces, such that f' corresponds to $\tilde{f}_{\tilde{b}}$ under i'.

By III.1.14 (iii), there exists a lifting $i': F' \to \tilde{E}$ of i such that the diagram

$$\begin{array}{ccc} F' & \xrightarrow{f'} & F' \\ {\scriptstyle i'}\downarrow & & \downarrow{\scriptstyle i'} \\ \tilde{E} & \xrightarrow{\tilde{f}} & \tilde{E} \end{array}$$

commutes. Now i' sends F' into some component $\tilde{F}_{\tilde{b}_1}$ of $p^{-1}(F_b)$, by Lemma 1.1 (ii). Obviously $\tilde{f}(\tilde{F}_{\tilde{b}}) \subset \tilde{F}_{\tilde{b}}$, hence $\tilde{b} \in \bar{p}^{-1}(b) \cap \mathrm{Fix}(\tilde{\tilde{f}})$. But $i': F' \to \tilde{F}_{\tilde{b}}$ is a lifting of the identity $F_b \to F_b$, and both F' and $\tilde{F}_{\tilde{b}}$ have the same K_b as fundamental group, so that i' is an isomorphism of covering spaces. □

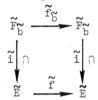

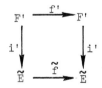

1.6 THEOREM. Let

$$\begin{array}{ccc} E & \xrightarrow{f} & E \\ {\scriptstyle q}\downarrow & & \downarrow{\scriptstyle q} \\ B & \xrightarrow{\bar{f}} & B \end{array} \qquad\qquad \begin{array}{ccc} \tilde{E} & \xrightarrow{\tilde{f}} & \tilde{E} \\ {\scriptstyle \tilde{q}}\downarrow & & \downarrow{\scriptstyle \tilde{q}} \\ \tilde{B} & \xrightarrow{\tilde{\bar{f}}} & \tilde{B} \end{array}$$

be a fiber map and a lifting of it.

(i) Suppose $b \notin \bar{p}\ \mathrm{Fix}(\tilde{\bar{f}})$. Then the fixed point class $p\ \mathrm{Fix}(\tilde{f})$ of f does not intersect the fiber F_b.

(ii) Suppose $b \in \bar{p}\ \mathrm{Fix}(\tilde{\bar{f}})$. Then the intersection of F_b with $p\ \mathrm{Fix}(\tilde{f})$ is a union of mod K_b fixed point classes of f_b, namely

$$F_b \cap p\ \mathrm{Fix}(\tilde{f}) = \bigcup_{\tilde{b} \in \bar{p}^{-1}(b) \cap \mathrm{Fix}(\tilde{\bar{f}})} p\ \mathrm{Fix}(\tilde{f}_{\tilde{b}}) .$$

The number of different mod K_b fixed point classes involved equals the index $[\mathrm{Fix}(\tilde{\bar{f}}_\pi) : \tilde{q}_\pi \mathrm{Fix}(\tilde{f}_\pi)]$.

(iii) The number $k_K(\tilde{f}) := [\mathrm{Fix}(\tilde{\bar{f}}_\pi) : \tilde{q}_\pi \mathrm{Fix}(\tilde{f}_\pi)]$ is determined by the homomorphism $\tilde{f}_\pi$. If $f_0 \simeq f_1$ is a homotopy of fiber maps which lifts to $\tilde{f}_0 \simeq \tilde{f}_1$, then $k_K(\tilde{f}_0) = k_K(\tilde{f}_1)$.

(iv) If $e \in p\ \mathrm{Fix}(\tilde{f})$, $b = q(e)$, then

$$k_K(\tilde{f}) = [\mathrm{Fix}(\pi_1(B,b) \xrightarrow{\bar{f}_\pi} \pi_1(B,b)) : q_\pi \mathrm{Fix}(\pi_1(E,e) \xrightarrow{f_\pi} \pi_1(E,e))] .$$

PROOF. (i) Trivial.

(ii) The decomposition is obvious. It remains to count the number of different mod K_b fixed point classes. By Lemma 1.5, this is equivalent to determining the cardinality $\# i_{FPC}^{-1}[\tilde{f}]$ for $i_{FPC} : FPC_{K_b}(f_b) \to FPC(f)$. Since $i : F_b \to E$ induces an injection $\pi_1(F_b)/K_b \to \pi_1(E)$, we may apply Proposition III.1.15 (see the end of §III.2) to conclude that $\# i_{FPC}^{-1}[\tilde{f}] = [\mathrm{Fix}(\tilde{f}_*) : \xi\ \mathrm{Fix}(\tilde{f}_\pi)]$, where $\xi : \pi_1(Y) \to \mathrm{Coker}\ \tilde{i}_\pi$ is the projection and $\tilde{f}_* : \mathrm{Coker}\ \tilde{i}_\pi \to \mathrm{Coker}\ \tilde{i}_\pi$ is induced by $\tilde{f}_\pi$.

By III.1.13 and the exactness of

$$\pi_1(F_b) \xrightarrow{i_\pi} \pi_1(E) \xrightarrow{q_\pi} \pi_1(B) \to 0 ,$$

we have the exactness of

$$\pi_1(F_b) \xrightarrow{\tilde{i}_\pi} \pi_1(E) \xrightarrow{\tilde{q}_\pi} \pi_1(B) \to 0 .$$

So we may identify $\mathrm{Coker}\ \tilde{i}_\pi$ with $\pi_1(B)$ via $\tilde{q}_\pi$. Then $\tilde{f}_*$ is identified with $\tilde{\bar{f}}_\pi$ since $\tilde{\bar{f}}_\pi \circ \tilde{q}_\pi = \tilde{q}_\pi \circ \tilde{f}_\pi$.

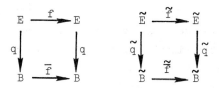

$$\begin{array}{ccc} \pi_1(\tilde E) & \xrightarrow{\tilde f_\pi} & \pi_1(\tilde E) \\ \downarrow \tilde q_\pi & & \downarrow \tilde q_\pi \\ \pi_1(\tilde B) & \xrightarrow{\tilde{\bar f}_\pi} & \pi_1(\tilde B) \end{array}$$

Hence $\# i_{FPC}^{-1}[\tilde f] = [\mathrm{Fix}(\tilde{\bar f}_\pi) : \tilde q_\pi \mathrm{Fix}(\tilde f_\pi)]$.

(iii) By the above commutative square, it is readily seen that the number $k_K(\tilde f)$ is completely determined by the homomorphism $\tilde f_\pi$. Then notice that $\tilde f_{0\pi} = \tilde f_{1\pi}$.

(iv) If we take $\tilde e \in p^{-1}(e) \cap \mathrm{Fix}(\tilde f)$ and $\tilde b = \tilde q(\tilde e) \in \bar p^{-1}(b) \cap \mathrm{Fix}(\tilde{\bar f})$ as base points in $\tilde E$ and $\tilde B$ respectively, then by III.1.13, we have $\tilde f_\pi = f_\pi$, $\tilde{\bar f}_\pi = \bar f_\pi$ and $\tilde q_\pi = q_\pi$. □

2. FIXED POINT CLASSES IN THE FIBER. From now on, we assume that $q : E \to B$ is a fibration with E, B and all fibers compact connected polyhedra, so that we can talk about the fixed point index in E, B and the fibers. We pick a regular lifting function once and for all, so that we can talk about fiber translations.

We will start with relatively simple results, and then refine our analysis.

It is well known that any two fibers F_b and $F_{b'}$ are of the same homotopy type. Let $f : E \to E$ be a fiber map, and let b and b' be two fixed points of $\bar f$. Are f_b and $f_{b'}$ necessarily of the same homotopy type (cf. Definition I.5.3)? The answer is "no" in general, as the following example shows.

EXAMPLE. Let K^2 be the Klein bottle. A fibration $q : K^2 \to S^1$ is shown in the figure, where K^2 is obtained by identifying the opposite sides of a rectangle as indicated. The fiber map $f : E \to E$ is a reflection with respect to the middle vertical line. It is obvious that $\mathrm{Fix}(\bar f) = \{b, b'\}$, and f_b has degree -1 while $f_{b'}$ has degree 1.

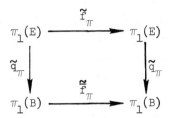

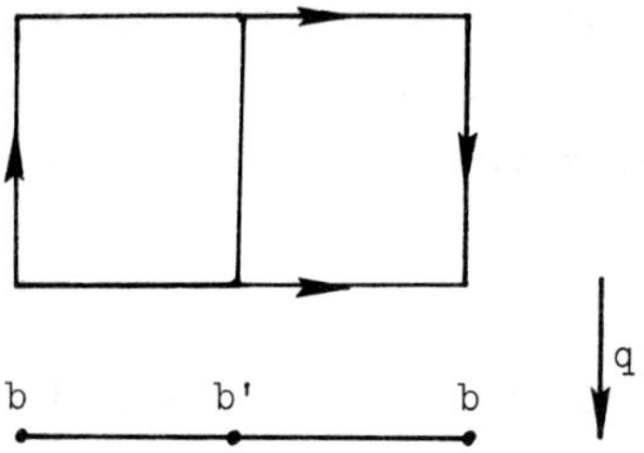

We may simplify the matter by introducing orientability.

2.1 DEFINITION. A fibration $q : E \to B$ is said to be (<u>homotopically</u>) <u>orientable</u> if for any two paths w, w' in B with the same endpoints $w(0) = w'(0)$ and $w(1) = w'(1)$, the fiber translations $\tau_w \simeq \tau_{w'} : F_{w(0)} \to F_{w(1)}$.

2.2 THEOREM. Suppose the fibration $q : E \to B$ is orientable. Let $f : E \to E$ be a fiber map. Then, for any two fixed points b, b' of $\bar{f} : B \to B$, the maps f_b and $f_{b'}$ have the same homotopy type; hence they have the same Lefschetz number, Nielsen number, and Reidemeister number.

Before the proof, let us introduce some notation and a lemma which will frequently be used.

2.3 NOTATION. For a path w and numbers $r, s \in I$, let w_r^s denote the path defined by $w_r^s(t) = w((1-t)r + ts)$. Thus w_1^0 is nothing but the reverse w^{-1} of w.

2.4 LEMMA. Let w be a path in B from b to b', both lying in $\mathrm{Fix}(\bar{f})$. Then

$$\tau_{\bar{f}\circ w} \circ f_b \simeq f_{b'} \circ \tau_w : F_b \to F_{b'} .$$

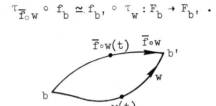

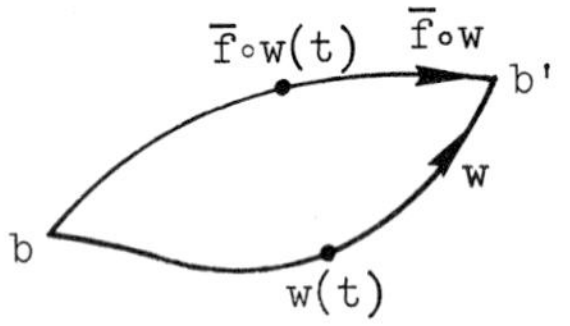

PROOF. Consider the homotopy $\{\tau_{\bar{f}\circ w_t^1} \circ (f \mid F_{w(t)}) \circ \tau_{w_0^t}\}_{t\in I}$. □

PROOF OF 2.2. By Lemma 2.4 and Definition 2.1, we see

$$\tau_w \circ f_b \simeq \tau_{\bar{f}\circ w} \circ f_b \simeq f_{b'} \circ \tau_w .$$

Notice that τ_w is a homotopy equivalence. □

Now comes a key observation: If we restrict our attention to two fixed points of $\bar{f}$ which are in the same fixed point class of $\bar{f}$, then no orientability assumption is necessary.

2.5 THEOREM. Let $f : E \to E$ be a fiber map inducing $\bar{f} : B \to B$, and let b and b' be in the same fixed point class of $\bar{f}$. Then f_b and $f_{b'}$ have the same Lefschetz number, Nielsen number and Reidemeister number. In fact they have the same homotopy type.

PROOF. Let $w : I \to B$ be a path from b to b' so that $w \simeq \bar{f} \circ w$. Then obviously $\tau_w \simeq \tau_{\bar{f}\circ w}$; hence by Lemma 2.4 we have

$$\tau_{\bar{f}\circ w} \circ f_b \simeq f_b \circ \tau_{\bar{f}\circ w} .$$ □

It is very instructive to put the proof in another form. Since $w \simeq \bar{f} \circ w$, there exists a continuous family of paths $\{c(t)\}_{t\in I}$ from b to b' such that $c(0) = \bar{f} \circ w$, $c(1) = w$. Let $g_1 = \tau_{\bar{f}\circ w^{-1}} \circ f_{b'} \circ \tau_w : F_b \to F_b$,

$g_2 = \tau_{w^{-1}} \circ f_{b'} \circ \tau_w : F_b \to F_b$, and $g_2' = \tau_w \circ \tau_{w^{-1}} \circ f_{b'} : F_{b'} \to F_{b'}$. Then $f_b \simeq g_1$ via the homotopy $\{\tau_{f \circ w_t^0} \circ (f \mid F_{w(t)}) \circ \tau_{w_0^t}\}_{t \in I}$, also $g_1 \simeq g_2$ via the homotopy $\{\tau_{c(t)^{-1}} \circ f_{b'} \circ \tau_w\}_{t \in I}$, and $g_2' \simeq f_{b'}$ via the homotopy $\{\tau_{w_t^1} \circ \tau_{w_1^t} \circ f_{b'}\}_{t \in I}$. And g_2' is obtained from g_2 by commuting the factors $(\tau_{w^{-1}} \circ f_{b'})$ and τ_w. Thus, by the homotopy invariance I.4.5 and the commutativity I.5.2 (or III.1.12), f_b and $f_{b'}$ have the same invariants.

The merit of this second proof is that it gives the construction of a one-one correspondence between the fixed point classes of f_b and those of $f_{b'}$. By "lifting" this argument to the universal coverings, we arrive at a significant result, a companion of Theorem 1.6.

2.6 THEOREM. Let the following be a fiber map and a lifting of it.

$$\begin{array}{ccc} E & \xrightarrow{f} & E \\ \downarrow q & & \downarrow q \\ B & \xrightarrow{\bar f} & B \end{array} \qquad \begin{array}{ccc} \tilde E & \xrightarrow{\tilde f} & \tilde E \\ \downarrow \tilde q & & \downarrow \tilde q \\ \tilde B & \xrightarrow{\tilde{\bar f}} & \tilde B \end{array}$$

Then the number

$$j_K(\tilde f) = \operatorname{index}(f_b, p\operatorname{Fix}(\tilde f_{\tilde b})) ,$$

where $b = \bar p(\tilde b)$, is independent of the point $\tilde b \in \operatorname{Fix}(\tilde{\bar f})$. Hence, given a fixed point class $p\operatorname{Fix}(\tilde f)$ of f, for any fiber F_b which intersects it, the intersection consists of $k_K(\tilde f)$ different mod K_b fixed point classes of f_b, each of which has the same index $j_K(\tilde f)$.

PROOF. Let $\tilde b, \tilde b' \in \operatorname{Fix}(\tilde{\bar f})$. Take a path $\tilde w : I \to \tilde B$ from $\tilde b$ to $\tilde b'$. Since $\tilde B$ is simply connected, there is a continuous family $\{\tilde c(t)\}_{t \in I}$ of paths from $\tilde b$ to $\tilde b'$ such that $\tilde c(0) = \tilde{\bar f} \circ \tilde w$ and $\tilde c(1) = \tilde w$. Let $w = \bar p \circ \tilde w$, etc. Let $\tilde g_1 = \tilde\tau_{\tilde{\bar f} \circ \tilde w^{-1}} \circ \tilde f_{\tilde b'} \circ \tilde\tau_{\tilde w} : \tilde F_{\tilde b} \to \tilde F_{\tilde b}$, let $\tilde g_2 = \tilde\tau_{\tilde w^{-1}} \circ \tilde f_{\tilde b'} \circ \tilde\tau_{\tilde w} : \tilde F_{\tilde b} \to \tilde F_{\tilde b}$, and let $\tilde g_2' = \tilde\tau_{\tilde w} \circ \tilde\tau_{\tilde w^{-1}} \circ \tilde f_{\tilde b'} : \tilde F_{\tilde b'} \to \tilde F_{\tilde b'}$. Then $\tilde g_1, \tilde g_2, \tilde g_2'$ are, respectively, liftings of the maps g_1, g_2, g_2' defined above. The homotopy $f_b \simeq g_1$ constructed before can be lifted to a homotopy $\{\tilde\tau_{\tilde{\bar f} \circ \tilde w_t^0} \circ (\tilde f \mid \tilde F_{\tilde w(t)}) \circ \tilde\tau_{\tilde w_0^t}\}_{t \in I} : \tilde F_{\tilde b} \to \tilde F_{\tilde b}$, and the homotopies $g_1 \simeq g_2$ and $g_2' \simeq f_{b'}$ can be similarly lifted. Thus, by the homotopy invariance III.2.5, we have

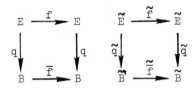

$$\mathrm{index}(f_b, p\,\mathrm{Fix}(\tilde f_{\tilde b})) = \mathrm{index}(g_2, p\,\mathrm{Fix}(\tilde g_2)) ,$$

$$\mathrm{index}(f_{b'}, p\,\mathrm{Fix}(\tilde f_{\tilde b'})) = \mathrm{index}(g_2', p\,\mathrm{Fix}(\tilde g_2')) .$$

But $\tilde g_2'$ is obtained from $\tilde g_2$ by commuting factors. By the commutativity III.2.6, we have

$$\mathrm{index}(g_2, p\,\mathrm{Fix}(\tilde g_2)) = \mathrm{index}(g_2', p\,\mathrm{Fix}(\tilde g_2')) .$$

Hence

$$\mathrm{index}(f_b, p\,\mathrm{Fix}(\tilde f_{\tilde b})) = \mathrm{index}(f_{b'}, p\,\mathrm{Fix}(\tilde f_{\tilde b'}) .$$

The second conclusion now follows from Theorem 1.6. □

2.7 REMARK. The definition of $j_{\tilde K}(\tilde f)$ given in the statement of Theorem 2.6 makes sense only if $\mathrm{Fix}(\tilde f)$ is non-empty. There is a general definition of the invariant $j_K(\tilde f)$, as an invariant of the lifting class $[\tilde f]$, as follows. Take any $\tilde b \in \tilde B$ and any path $\tilde c$ from $\tilde{\bar f}(\tilde b)$ to $\tilde b$, let $b = \bar p(\tilde b)$, $c = \bar p \circ \tilde c$, and define

$$j_K(\tilde f) = \mathrm{index}(\tau_c \circ (f \mid F_b), p\,\mathrm{Fix}(\tilde\tau_{\tilde c} \circ (\tilde f \mid \tilde F_{\tilde b})) .$$

The independence of the choice of $\tilde b$ and $\tilde c$ can be proved in much the same way as above.

The same remark applies to the definition of $k'(\tilde f)$, $k(\tilde f)$ and $j(\tilde f)$ in Theorem 2.11.

Our next step is to investigate the ordinary fixed point classes in a fiber. It turns out that the cyclic homotopies play a key role.

2.8 THEOREM. For any $b \in B$, the normal subgroup $K_b = \mathrm{Ker}(i_\pi : \pi_1(F_b) \to \pi_1(E))$ is contained in $J(F_b)$.

PROOF. Take a point $e \in E$ as base point for F_b and E. By the homotopy exact sequence of the fibration $q : E \to B$, we know $K_b = \mathrm{Im}(\partial : \pi_2(B,b) \to \pi_1(F_b,e))$. Let $[\bar\varphi] \in \pi_2(B,b)$ be an element represented by a map $\bar\varphi : I^2, \dot I^2 \to B, b$. We want to show $\partial[\bar\varphi] \in J(F_b,e)$.

By definition, $\partial[\bar\varphi] = [\varphi_1]$, where $\varphi : I^2, \dot I^2, I \times \dot I \cup 0 \times I \to E, F_b, e$ is any map with $q \circ \varphi = \bar\varphi$, and $\varphi_1 : I, \dot I \to F_b, e$ is obtained from φ by $\varphi_1(t) = \varphi(1,t)$.

We construct a φ as follows. For $s, t \in I$, let $c(s,t)$ be the path in B defined by $c(s,t)(u) = \bar\varphi(su,t)$, $u \in I$. Let $\varphi(s,t) = \tau_{c(s,t)}(e)$. Obviously, $q \circ \varphi = \bar\varphi$, hence $\partial[\bar\varphi]$ is represented by the path $\{\varphi_1(t)\}_{t\in I} = \{\tau_{c(1,t)}(e)\}_{t\in I}$, that is, the trace of the homotopy

$\{\tau_{c(1,t)}\}_{t\in I}$. But this is a cyclic homotopy $id \simeq id : F_b \to F_b$, since $c(1,0)$ and $c(1,1)$ are both constant paths. Thus $\partial[\bar{\varphi}] \in J(F_b, e)$. □

2.9 COROLLARY. K_b is contained in the center of $\pi_1(F_b)$.

PROOF. Apply II.3.7. □

2.10 COROLLARY. Any two ordinary fixed point classes in F_b which are contained in the same mod K_b fixed point class have the same index.

PROOF. Apply III.2.11. □

We are now well prepared for another significant result.

2.11 THEOREM. The same hypothesis as in Theorem 2.6. Then, the number $k'(\tilde{f})$ of ordinary fixed point classes of f_b which are contained in $p\,\mathrm{Fix}(\tilde{f}_{\tilde{b}})$, and the index $j(\tilde{f})$ of each such ordinary fixed point class, are independent of the point $\tilde{b} \in \mathrm{Fix}(\tilde{\bar{f}})$, i.e., are completely determined by the lifting class $[\tilde{f}]$. Thus, given a fixed point class $p\,\mathrm{Fix}(\tilde{f})$ of f, for any fiber F_b that intersects it (that is, F_b with $b \in \bar{p}\,\mathrm{Fix}(\tilde{\bar{f}})$), the intersection consists of $k(\tilde{f}) = k_K(\tilde{f}) \cdot k'(\tilde{f})$ different fixed point classes of f_b, each of which has the same index $j(\tilde{f})$. In particular, $j(\tilde{f}) \neq 0$ implies $k(\tilde{f}) < \infty$ and $q(p\,\mathrm{Fix}(\tilde{f})) = \bar{p}\,\mathrm{Fix}(\tilde{\bar{f}})$.

PROOF. The construction given in the proof of 2.6 establishes a correspondence, one-to-one and index-preserving, between ordinary fixed point classes of f_b and ordinary fixed point classes of $f_{b'}$, which makes the mod K_b fixed point class $p\,\mathrm{Fix}(\tilde{f}_{\tilde{b}})$ correspond to the mod $K_{b'}$ fixed point class $p\,\mathrm{Fix}(\tilde{f}_{\tilde{b}'})$. Hence the first conclusion. Now combine with Theorem 1.6 (ii) to get the second conclusion. The last conclusion follows trivially. □

2.12 REMARK. $j_K(\tilde{f}) = k'(\tilde{f})j(\tilde{f})$.

2.13 THEOREM. If $\{f_t\}_{t\in I} : f_0 \simeq f_1 : E \to E$ is a homotopy of fiber maps which lifts to $\{\tilde{f}_t\}_{t\in I} : \tilde{f}_0 \simeq \tilde{f}_1 : \tilde{E} \to \tilde{E}$, then $k(\tilde{f}_0) = k(\tilde{f}_1)$ and $j(\tilde{f}_0) = j(\tilde{f}_1)$. In other words, $k(\tilde{f})$ and $j(\tilde{f})$, as well as $k_K(\tilde{f})$, $k'(\tilde{f})$ and $j_K(\tilde{f})$, are invariant under a homotopy of fiber maps.

PROOF. Consider the fibration $q \times id : E \times I \to B \times I$. The fiber homotopy $\{f_t\}$ gives rise to a "fat homotopy" (cf. I.2.5) $\mathbf{f} : E \times I \to E \times I$ which is a fiber map with respect to this new fibration. Then apply 2.11 to this new fiber map. □

We have already had a way of computing $k_K(\tilde{f})$, namely Theorem 1.6 (iv). The following is a way of computing $k'(\tilde{f})$.

2.14 PROPOSITION. If $e \in p\,\mathrm{Fix}(\tilde{f})$ and $b = q(e)$, then

$$k'(\tilde{f}) = [K_b : L_b] ,$$

where L_b is the subgroup of K_b consisting of elements that are expressible as $\gamma f_{b\pi}(\gamma^{-1})$ for some $\gamma \in \pi_1(F_b, e)$.

PROOF. This follows from the definition of $k'(\tilde{f})$ in 2.11, and III.2.11. □

3. ESSENTIAL FIXED POINT CLASSES IN THE TOTAL SPACE. The following lemma is very useful in computing the fixed point index in the total space E.

3.1 LEMMA. Let $q : E \to B$ be a fibration and let $f : E \to E$ be a fiber map inducing $\bar{f} : B \to B$. Suppose b_0 is an isolated fixed point of $\bar{f}$, and let $f_0 = f \mid F_{b_0}$. If $A \subset \mathrm{Fix}(f_0)$ is an isolated set of fixed points of f_0 (see Definition I.3.8), then A is also an isolated set of fixed points of f, and

$$\mathrm{index}(f,A) = \mathrm{index}(\bar{f},b_0) \cdot \mathrm{index}(f_0,A) .$$

PROOF. To simplify notation, embed B linearly in some Euclidean space so that $b_0 = 0$. Let S be a star-shaped neighborhood of 0, that is, every point of S can be joined to 0 by a straight line segment. Let U be a neighborhood of 0 such that $\bar{U} \cup \overline{f(U)} \subset S$ and 0 is the only fixed point on $\bar{U}$. For any two points $b, b' \in S$ which can be joined by a linear path, let $\tau_{b,b'} : F_b \to F_{b'}$ be the fiber translation determined by that linear path. Note that $\tau_{b,b} = \mathrm{id}$.

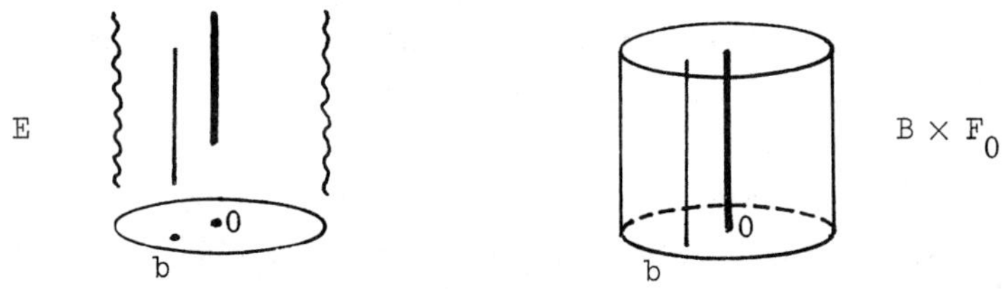

Define two maps

$$\varphi : q^{-1}(S) \to B \times F_0, \qquad e \mapsto (q(e), \tau_{q(e),0}(e)) ,$$

$$\psi : U \times F_0 \to E, \qquad (b,e_0) \mapsto f \circ \tau_{0,b}(e_0) .$$

There is a homotopy $\{h_t\}_{t\in I} : \psi \circ \varphi \simeq f : q^{-1}(U) \to E$ defined by

$$h_t : e \mapsto f \circ \tau_{tq(e),q(e)} \circ \tau_{q(e),tq(e)}(e) ,$$

and a homotopy $\{k_t\}_{t\in I} : \bar{f} \times f_0 \simeq \varphi \circ \psi : U \times F_0 \to B \times F_0$ defined by

$$k_t : (b,e_0) \mapsto (\bar{f}(b), \tau_{\bar{f}(tb),0} \circ f \circ \tau_{0,tb}(e_0)) .$$

The continuity of $\{h_t\}$ and $\{k_t\}$ are obvious, and the fixed point set of h_t and k_t are independent of $t \in I$.

If $A \subset \mathrm{Fix}(f_0)$ is an isolated set of fixed points of f_0, it is also an isolated set of fixed points of f since b_0 is an isolated point of $\bar{f}$. We have

$$\begin{aligned}
\mathrm{index}(f,A) &= \mathrm{index}(\psi \circ \varphi, A) && \text{by I.3.5 (ii)}\\
&= \mathrm{index}(\varphi \circ \psi, b_0 \times A) && \text{by I.3.5 (v)}\\
&= \mathrm{index}(\bar{f} \times f_0, b_0 \times A) && \text{by I.3.5 (ii)}\\
&= \mathrm{index}(f, b_0) \cdot \mathrm{index}(f_0, A) && \text{by I.3.5 (iv)}. \qquad \square
\end{aligned}$$

Our first application of this lemma is:

3.2 THEOREM (Product formula for the Lefschetz number). Let $q : E \to B$ be a fibration. Let $f : E \to E$ be a fiber map inducing $\bar{f} : B \to B$ such that $L(f_b)$ is the same for all fixed points $b \in \mathrm{Fix}(\bar{f})$. Then

$$L(f) = L(\bar{f}) \cdot L(f_b) .$$

PROOF. Without loss of generality, we may assume $\bar{f}$ has only isolated fixed points. (Otherwise we may homotope $\bar{f}$ to $\bar{g}$ having only isolated fixed points, by the Hopf Approximation Theorem, and then lift this homotopy to a homotopy $f \simeq g : E \to E$ by the homotopy lifting property of the fibration.) Let $\{b_i\}$ be the fixed points of $\bar{f}$. Now

$$\begin{aligned}
L(f) &= \mathrm{index}(f, \textstyle\bigcup_i \mathrm{Fix}(f_{b_i}))\\
&= \sum_i \mathrm{index}(f, \mathrm{Fix}(f_{b_i}))\\
&= \sum_i \mathrm{index}(\bar{f}, b_i) \cdot \mathrm{index}(f_{b_i}, \mathrm{Fix}(f_{b_i})) && \text{(by Lemma 3.1)}\\
&= \sum_i \mathrm{index}(\bar{f}, b_i) \cdot L(f_{b_i})\\
&= (\sum_i \mathrm{index}(\bar{f}, b_i)) \cdot L(f_b) && \text{(by the assumption that the } L(f_{b_i}) \text{ are the same)}\\
&= L(\bar{f}) \cdot L(f_b) . \qquad \square
\end{aligned}$$

EXERCISE. If $\bar{f} = \mathrm{id} : B \to B$, then $L(f_b)$ is independent of $b \in B$, hence $L(f) = L(\bar{f})L(f_b) = \chi(B) \cdot L(f_b)$.

3.3 THEOREM (Product formula for the index of a fixed point class). Let $q : E \to B$ be a fibration and let $f : E \to E$ be a fiber map inducing $\bar{f} : B \to B$. Let $\tilde{f} : \tilde{E} \to \tilde{E}$ be a lifting of f inducing $\tilde{\bar{f}} : \tilde{B} \to \tilde{B}$. Then for any $b \in \bar{p}\,\mathrm{Fix}(\tilde{\bar{f}})$, we have

$$\mathrm{index}(f, p\,\mathrm{Fix}(\tilde{f})) = \mathrm{index}(\bar{f}, \bar{p}\,\mathrm{Fix}(\tilde{\bar{f}})) \cdot \mathrm{index}(f_b, F_b \cap p\,\mathrm{Fix}(\tilde{f})) .$$

PROOF. From Theorem 2.6 we know $\operatorname{index}(f_b, F_b \cap p\operatorname{Fix}(\tilde{f}))$ is independent of $b \in \bar{p}\operatorname{Fix}(\tilde{\bar{f}})$.

Without loss of generality we may assume $\bar{f}$ has only isolated fixed points. Otherwise we may apply the Hopf Approximation Theorem to get $\bar{f} \simeq \bar{g} : B \to B$ such that $\bar{g}$ has only isolated fixed points, and we may lift this homotopy to $f \simeq g : E \to E$. But by the homotopy invariance of the index of fixed point classes we have

$$\operatorname{index}(f, p\operatorname{Fix}(\tilde{f})) = \operatorname{index}(g, p\operatorname{Fix}(\tilde{g})) ,$$

$$\operatorname{index}(\bar{f}, p\operatorname{Fix}(\tilde{\bar{f}})) = \operatorname{index}(\bar{g}, \bar{p}\operatorname{Fix}(\tilde{\bar{g}})) ,$$

and by Theorem 2.13 we have

$$\operatorname{index}(f_b, F_b \cap p\operatorname{Fix}(\tilde{f})) = \operatorname{index}(g_{b'}, F_{b'} \cap p\operatorname{Fix}(\tilde{g}))$$

for $b \in \bar{p}\operatorname{Fix}(\tilde{\bar{f}})$ and $b' \in \bar{p}\operatorname{Fix}(\tilde{\bar{g}})$. So the product formula for f and for g are equivalent.

Now let $\{b_i\}$ be the (isolated) fixed points in $\bar{p}\operatorname{Fix}(\tilde{\bar{f}})$. We have

$$\begin{aligned}
\operatorname{index}(f, p\operatorname{Fix}(\tilde{f})) &= \operatorname{index}(f, \bigcup_i F_{b_i} \cap p\operatorname{Fix}(\tilde{f})) \\
&= \sum_i \operatorname{index}(f, F_{b_i} \cap p\operatorname{Fix}(\tilde{f})) \\
&= \sum_i \operatorname{index}(\bar{f}, b_i) \cdot \operatorname{index}(f_{b_i}, F_{b_i} \cap p\operatorname{Fix}(\tilde{f})) \text{ (by Lemma 3.1)} \\
&= (\sum_i \operatorname{index}(\bar{f}, b_i)) \cdot \operatorname{index}(f_b, F_b \cap p\operatorname{Fix}(\tilde{f})) \\
&= \operatorname{index}(\bar{f}, \bar{p}\operatorname{Fix}(\tilde{\bar{f}})) \cdot \operatorname{index}(f_b, F_b \cap p\operatorname{Fix}(\tilde{f})) . \qquad \square
\end{aligned}$$

3.4 THEOREM. The fixed point class $p\operatorname{Fix}(\tilde{f})$ is essential iff $\bar{p}\operatorname{Fix}(\tilde{\bar{f}})$ is essential and $j'(\tilde{f}) \neq 0$. In other words, a fixed point class of f is essential iff its projection in B is an essential fixed point class of $\bar{f}$ and its intersection with an invariant fiber F_b consists of essential fixed point classes of f_b. □

3.5 REMARK. In view of Theorem 3.4, we can see that in Theorem 3.2, what is important is $L(f_b)$ for b in essential fixed point classes of $\bar{f}$. So we may improve Theorem 3.2 to

3.6 THEOREM. Suppose $f : E \to E$ is a fiber map inducing $\bar{f} : B \to B$, such that $L(f_b)$ is the same for any b in any essential fixed point class of $\bar{f}$. Then, for any b in any essential fixed point class of $\bar{f}$, we have

$$L(f) = L(\bar{f}) \cdot L(f_b) . \qquad \square$$

For example, if $N(\bar{f}) \leq 1$, then this remark applies.

4. PRODUCT FORMULAS FOR THE NIELSEN NUMBER. Inspired by Theorem 3.2, we would naturally ask whether, if a fiber map $f : E \to E$ is such that $N(f_b)$ is the same for any $b \in Fix(\bar{f})$, it follows that $N(f) = N(\bar{f}) \cdot N(f_b)$. This striking example is due to Fadell.

EXAMPLE. Let $S^1 \to S^3 \xrightarrow{q} S^2$ be the Hopf fibration. Then, if $\bar{f} : S^2 \to S^2$ is a map of degree d, there is no obstruction to lifting $\bar{f}$ to a fiber map $f : S^3 \to S^3$. Suppose $d \neq -1$ so that $\bar{f}$ has a fixed point $b \in S^2$. It is easy to see that f_b also has degree d. Then, computing Lefschetz numbers we obtain $L(\bar{f}) = 1 + d$, $L(f_b) = 1 - d$, hence by Theorem 3.2 $L(f) = 1 - d^2$. Since S^3 and S^2 are simply connected, it is clear that $N(f) = N(\bar{f}) = 1$ as long as $d \neq \pm 1$. But by Theorem II.4.1, $N(f_b) = |1 - d|$. Thus, if we take $|d| \geq 3$, we have $N(f) \neq N(\bar{f}) \cdot N(f_b)$.

Now let us use all the information from previous sections to analyze $N(f)$. Let $\bar{\mathbb{F}}_i$, $i = 1,2,\ldots,N(\bar{f})$, be the essential fixed point classes of $\bar{f}$. According to Theorem 3.4, each essential fixed point class of f projects to an essential fixed point class of $\bar{f}$. Let $\mathbb{F}_{i1},\ldots,\mathbb{F}_{ic_i}$ be the essential fixed point classes of f lying above $\bar{\mathbb{F}}_i$. Let k_{ij} (resp. k_{ijK}) be the number of (essential) ordinary (resp. mod K_b) fixed point classes of f_b contained in $F_b \cap \mathbb{F}_{ij}$, where $b \in \bar{\mathbb{F}}_i$. These numbers are independent of $b \in \bar{\mathbb{F}}_i$ by Theorem 2.11. Thus

$$N(f) = \sum_{i=1}^{N(\bar{f})} c_i \,,$$

$$N(f_b) = \sum_{j=1}^{c_i} k_{ij}, \qquad \text{if } b \in \bar{\mathbb{F}}_i \,,$$

$$N_K(f_b) = \sum_{j=1}^{c_i} k_{ijK}, \qquad \text{if } b \in \bar{\mathbb{F}}_i \,.$$

To understand the relationship between $N(f)$ and $N(\bar{f}) \cdot N(f_b)$ or $N(\bar{f}) \cdot N_K(f_b)$, it is crucial to have information about k_{ij} or k_{ijK}, i.e. how many ordinary or mod K_b fixed point classes of $f_b : F_b \to F_b$ are contained in a fixed point class $\mathbb{F}_{ij}$ of f.

<u>The naïve product formula</u> $N(f) = N(\bar{f}) \cdot N(f_b)$. Let us assume $N(f_b)$ is independent of b in essential fixed point classes of $\bar{f}$. (This is the case if the fibration is orientable, or $N(\bar{f}) = 1$.) By the above analysis,

$$N(\bar{f}) \cdot N(f_b) = \sum_{i=1}^{N(\bar{f})} \sum_{j=1}^{c_i} k_{ij} \geq \sum_{i=1}^{N(\bar{f})} \sum_{j=1}^{c_i} k_{ijK} \geq \sum_{i=1}^{N(\bar{f})} \sum_{j=1}^{c_i} 1 = \sum_{i=1}^{N(\bar{f})} c_i = N(f) \,,$$

since $k_{ij} \geq k_{ijK} \geq 1$. So, the naïve product formula is true iff every $k_{ij} = k_{ijK} = 1$.

4.1 THEOREM. Let $q : E \to B$ be a fibration. Let $f : E \to E$ be a fiber map such that $N(f_b)$ is independent of b in essential fixed point classes of $\bar{f}$. Then the naïve product formula $N(f) = N(\bar{f}) \cdot N(f_b)$ holds iff

(a) $N_K(f_b) = N(f_b)$, where $K = K_b = \mathrm{Ker}(i_\pi : \pi_1(F_b) \to \pi_1(E))$, and

(b) in every essential fixed point class of f, there is a point e such that

$$q_\pi \mathrm{Fix}(\pi_1(E,e) \xrightarrow{f_\pi} \pi_1(E,e)) = \mathrm{Fix}(\pi_1(B,b) \xrightarrow{\bar{f}_\pi} \pi_1(B,b)) ,$$

where $b = q(e)$.

PROOF. Condition (a) is evidently equivalent to saying that $k_{ij} = k_{ijK}$. For $e \in \mathbb{F}_{ij}$, condition (b) is equivalent to $k_{ijK} = 1$ by Theorem 1.6. □

4.2 THEOREM. Let $q : E \to B$ be a fibration. Let $f : E \to E$ be a fiber map such that $N(f_b)$ is independent of b in essential fixed point classes of $\bar{f}$. Then, the naïve product formula $N(f) = N(\bar{f}) \cdot N(f_b)$ holds iff in every essential fixed point class of f, there is a point e such that

(a) every element in $K_b = \mathrm{Ker}(\pi_1(F_b,e) \xrightarrow{i_\pi} \pi_1(E,e))$ equals $\gamma f_{b\pi}(\gamma^{-1})$ for some $\gamma \in \pi_1(F_b,e)$, and

(b) $q_\pi(\mathrm{Fix}(\pi_1(E,e) \xrightarrow{f_\pi} \pi_1(E,e))) = \mathrm{Fix}(\pi_1(B,b) \xrightarrow{\bar{f}_\pi} \pi_1(B,b))$, where $b = q(e)$.

PROOF. For $e \in \mathbb{F}_{ij}$, condition (a) is equivalent to $k_{ij} = k_{ijK}$ by Theorems 2.11 and 2.14, while (b) is equivalent to $k_{ijK} = 1$ by Theorem 1.6. □

Theorems 4.1 and 4.2 have a number of corollaries.

4.3 COROLLARY. The naïve product formula holds if one of the following conditions is satisfied:

(i) $N(\bar{f}) = 0$,

(ii) $N(f_b) \leq 1$ for any $b \in \bar{\mathbb{F}}_i$, $i = 1,\dots,N(\bar{f})$. □

4.4 COROLLARY. Suppose $q : E \to B$ is an orientable fibration and $i_\pi : \pi_1(F_b) \to \pi_1(E)$ is injective. Suppose $f : E \to E$ is a fiber map such that the fixed element subgroup $\mathrm{Fix}(\pi_1(B,b) \xrightarrow{\bar{f}_\pi} \pi_1(B,b))$ is trivial for some b in every essential fixed point class of $\bar{f}$. Then $N(f) = N(\bar{f}) \cdot N(f_b)$. □

4.5 COROLLARY. Suppose $q : E \to B$ is an orientable fibration, $f : E \to E$ is a fiber map. Suppose they admit a Fadell splitting in the sense that for some $e \in \mathrm{Fix}(f)$ and $b = q(e)$ the following conditions are satisfied:

(a) the sequence

$$0 \to \pi_1(F_b,e) \xrightarrow{i_\pi} \pi_1(E,e) \xrightarrow{q_\pi} \pi_1(B,b) \to 0$$

is exact; and

(b) q_π admits a right inverse (section) σ such that $\operatorname{Im}\sigma \lhd \pi_1(E,e)$ and $f_\pi(\operatorname{Im}\sigma) \subset \operatorname{Im}\sigma$.

Then the naïve product formula holds.

PROOF. We first show that if a Fadell splitting exists for one $e \in \operatorname{Fix}(f)$ then it exists for any $e \in \operatorname{Fix}(f)$. In fact, let e' be another fixed point and w be a path from e to e', then $\bar{w} = q \circ w$ is a path from b to b', and we may define σ' by $\sigma' = w_*^{-1} \circ \sigma \circ \bar{w}_*$. Obviously

$$\begin{array}{ccccc}
\pi_1(B,b) & \xrightarrow{\ \sigma\ } & \pi_1(E,e) & \xrightarrow{\ q_\pi\ } & \pi_1(B,b) \\
\uparrow{\scriptstyle \bar{w}_*} & & \uparrow{\scriptstyle w_*} & & \uparrow \\
\pi_1(B,b') & \xrightarrow{\ \sigma'\ } & \pi_1(E,e') & \xrightarrow{\ q_\pi\ } & \pi_1(B,b')
\end{array}$$

and σ' is also a Fadell splitting.

Condition (a) implies K_b is trivial and (a) of Theorem 4.2 is satisfied. Let us verify (b) of 4.2. For $\bar\alpha \in \operatorname{Fix}(\bar f_\pi)$, let $\alpha = \sigma(\bar\alpha)$. We have $f_\pi(\alpha) = \sigma(\bar\beta)$ for some $\bar\beta \in \pi_1(B,b)$. Then $\bar\beta = q_\pi \circ f_\pi(\alpha) = \bar f_\pi \circ q_\pi(\alpha)$ $= \bar f_\pi(\bar\alpha) = \bar\alpha$, so $f_\pi(\alpha) = \sigma(\bar\alpha) = \alpha$. Thus $\bar\alpha \in q_\pi \operatorname{Fix}(f_\pi)$. □

The merit of the Fadell splitting is that we only have to check the condition at one fixed point.

4.6 COROLLARY. Suppose the fiber map $f : E \to E$ is such that the homomorphism $1 - \bar f_\pi : \pi_2(B,b) \to \pi_2(B,b)$ is an epimorphism. Then the condition (a) of Theorem 4.2 is always satisfied.

PROOF. Let $e \in \operatorname{Fix}(f)$ and $b = q(e)$. Check the following commutative diagram:

$$\begin{array}{ccc}
\pi_2(B,b) & \xrightarrow{\ \partial\ } & \pi_1(F_b,e) \\
\downarrow{\scriptstyle \bar f_\pi} & & \downarrow{\scriptstyle f_{b\pi}} \\
\pi_2(B,b) & \xrightarrow{\ \partial\ } & \pi_1(F_b,e)
\end{array},$$

where $K_b = \operatorname{Im}\partial$. Then, each element $\partial\zeta$ of K_b, $\zeta \in \pi_2(B,b)$, with $\zeta = \xi - \bar f_\pi(\xi)$ by assumption, can be written as

$$\partial\zeta = \partial(\xi - \bar f_\pi(\xi)) = (\partial\xi)(\partial\bar f_\pi(\xi))^{-1} = (\partial\xi)(f_{b\pi}\partial\xi)^{-1} = \gamma f_{b\pi}(\gamma^{-1})$$

where $\gamma = \partial\xi$. □

The following example shows how the above conditions can be put to use.

EXAMPLE. Let $q : E \to B$ be orientable and $B = T^n$, the n-torus. Then the naïve product formula holds for any fiber map $f : E \to E$.

In fact, $\pi_2(B) = 0$ implies $K_b = 0$, hence (a) of Theorem 4.2 holds. On the other hand, Example 2 of §II.4 tells us that $N(\bar{f}) > 0$ iff $\mathrm{Fix}(\bar{f}_\pi)$ is trivial. Now the conclusion follows from Corollary 4.4.

The following two exercises indicate a way of constructing simple examples (and counterexamples).

EXERCISE 1. Consider the trivial circle bundle over the figure eight. Namely, let $B = \{z \in \mathbb{C} \mid |z| = 1 \text{ or } |z - 2| = 1\}$, $F = \{w \in \mathbb{C} \mid |w| = 1\}$, and $E = B \times F$. A fiber map $f : E \to E$ is defined by

$$f(z,w) = \begin{cases} (z^n, z^\ell w^m) & \text{if } |z| = 1 , \\ (z, w^m) & \text{if } |z - 2| = 1 , \end{cases}$$

where ℓ, m, n are integers, $n \neq 1$. Then the naïve product formula holds by Theorem 4.2. Note that when $(n - m) \nmid \ell$ it does not satisfy the condition of 4.4 and 4.5.

EXERCISE 2. The fibration is the same as above. The fiber map $f : E \to E$ is defined by

$$f(z,w) = \begin{cases} (z^3, w^{-1}) & \text{if } |z| = 1 , \\ (z, -(z - 2)w^{-1}) & \text{if } |z - 2| = 1 . \end{cases}$$

Then $N(\bar{f}) = 2$, $N(f_b) = 2$, while $N(f) = 3$. Note that the condition (b) of Theorem 4.2 is satisfied at the fixed points $(-1,1)$ and $(-1,-1)$, but not at other fixed points.

Product formula with a correction factor. By the general analysis at the beginning of this section, for an orientable fibration we always have $N(\bar{f}) \cdot N(f_b) \geq N(f)$, so, when $N(f) \neq 0$, we have the rational number $N(\bar{f}) \cdot N(f_b)/N(f) \geq 1$. In some cases this factor is explicitly computable.

4.7 DEFINITION. Suppose we have a commutative square of maps

$$\begin{array}{ccc} X & \xrightarrow{f} & X \\ {\scriptstyle h}\downarrow & & \downarrow{\scriptstyle h} \\ Y & \xrightarrow{g} & Y \end{array} \qquad (*)$$

For integral homology there is an induced diagram

$$\begin{array}{ccc} H_1(X) & \xrightarrow{1-f_*} & H_1(X) \\ {\scriptstyle h_*}\downarrow & & \downarrow{\scriptstyle h_*} \\ H_1(Y) & \xrightarrow{1-g_*} & H_1(Y) \end{array}$$

hence h_* induces a homomorphism

$$h_*' : \mathrm{Coker}(1 - f_*) \to \mathrm{Coker}(1 - g_*) .$$

The Pak number of the square (*) is defined to be the order of $\mathrm{Ker}\ h_*'$.

4.8 LEMMA. Let X, Y be compact connected polyhedra. Let

$$\begin{array}{ccc} X & \xrightarrow{f} & X \\ {\scriptstyle h}\downarrow & & \downarrow{\scriptstyle h} \\ Y & \xrightarrow{g} & Y \end{array} \qquad (*)$$

be a commutative square in which f and g are eventually commutative (cf. II.2.4). Consider a fixed point class of g. Assume there exist fixed point classes of f which are mapped by h into this fixed point class of g. Then, the number of such fixed point classes of f equals the Pak number of the square (*).

PROOF. The fixed point classes of f are in one-one correspondence with the elements of $\mathrm{Coker}(1 - f_{1*})$, according to Theorem II.2.5; if we specify a lifting $\tilde{f}$ of f as reference, then the fixed point class $p_X\mathrm{Fix}(\alpha \circ \tilde{f})$, $\alpha \in \pi_1(X)$, corresponds to $\eta \circ \theta(\alpha)$, where $\theta : \pi_1(X) \to H_1(X)$ is the abelianization, $\eta : H_1(X) \to \mathrm{Coker}(1 - f_*)$ is the natural projection. Similarly for g.

Pick a commutative lifting of (*),

$$\begin{array}{ccc} \tilde{X} & \xrightarrow{\tilde{f}} & \tilde{X} \\ {\scriptstyle \tilde{h}}\downarrow & & \downarrow{\scriptstyle \tilde{h}} \\ \tilde{Y} & \xrightarrow{\tilde{g}} & \tilde{Y} \end{array}$$

as reference, Then, for $\alpha \in \pi_1(X)$, we have $\check{h} \circ \alpha = \tilde{h}_\pi(\alpha) \circ \tilde{h}$, hence

$$\begin{array}{ccc} \tilde{X} & \xrightarrow{\alpha\circ\tilde{f}} & \tilde{X} \\ {\scriptstyle\tilde{h}}\downarrow & & \downarrow{\scriptstyle\tilde{h}} \\ \tilde{Y} & \xrightarrow{\tilde{h}_\pi(\alpha)\ \tilde{g}} & \tilde{Y} \end{array}$$

commutes. Thus, the fixed point class $p_X \mathrm{Fix}(\alpha \circ \tilde{f})$ is mapped by h into the fixed point class $p_Y \mathrm{Fix}(\tilde{h}_\pi(\alpha) \circ \tilde{g})$.

Now the fixed point class of f corresponding to $\eta \circ \theta(\alpha) \in \mathrm{Coker}(1 - f_*)$ is mapped by h into the fixed point class of g corresponding to $\eta \circ \theta(\tilde{h}_\pi(\alpha)) = \eta \circ h_*(\theta(\alpha)) = H'_*(\eta \circ \theta(\alpha)) \in \mathrm{Coker}(1 - g_*)$. But h'_* is a homomorphism, so the Pak number tells us how many elements of $\mathrm{Coker}(1 - f_*)$ are sent into the same element of $\mathrm{Coker}(1 - g_*)$. □

Suppose $f : E \to E$ is a fiber map, $b \in \mathrm{Fix}(\bar{f})$. Then we have a commutative square

$$\begin{array}{ccc} F_b & \xrightarrow{f_b} & F_b \\ {\scriptstyle i}\downarrow & & \downarrow{\scriptstyle i} \\ E & \xrightarrow{f} & E \end{array} \qquad (*)_b$$

4.9 LEMMA. (i) If f_b is eventually commutative for some $b \in \mathrm{Fix}(\bar{f})$, then it is so for all $b \in \mathrm{Fix}(\bar{f})$.

(ii) Suppose the fibration is orientable. Then the Pak number for the square $(*)_b$ is independent of $b \in \mathrm{Fix}(\bar{f})$.

PROOF. (i) Let $b, b' \in \mathrm{Fix}(\bar{f})$, and let c be a path from b to b'. By Lemma 2.4,

$$f^k_{b'} \circ \tau_c \simeq \tau_{f^k c} \circ f^k_b .$$

Since τ_c and $\tau_{f^k \circ c}$ induce isomorphisms on π_1, $(f^k_b)_\pi \pi_1(F_b)$ is commutative iff $(f^k_{b'})_\pi \pi_1(F_{b'})$ is commutative.

(ii) Consider the diagram

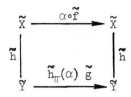

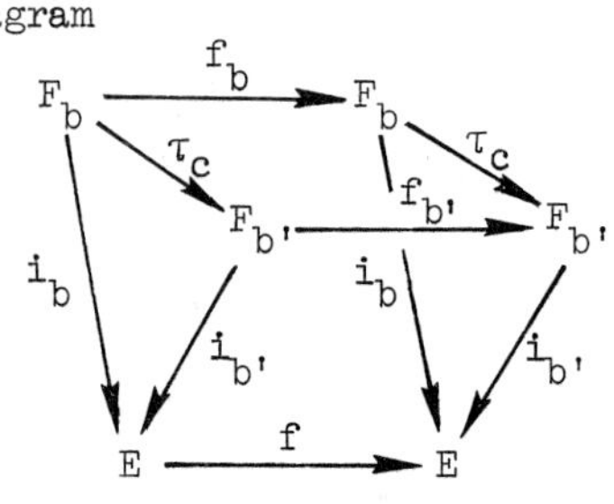

This diagram is homotopically commutative, since the upper parallelogram is homotopically commutative as shown in Lemma 2.2. Our conclusion follows from the induced diagram in homology. □

4.10 THEOREM. Let $q : E \to B$ be orientable. Let $f : E \to E$ be a fiber map such that f and $f_b : F_b \to F_b$ are eventually commutative for $b \in \mathrm{Fix}(\bar{f})$. Let $P(f)$ be the Pak number of the square

$$\begin{array}{ccc} F_b & \xrightarrow{f_b} & F_b \\ {\scriptstyle i_b}\downarrow & & \downarrow{\scriptstyle i_b} \\ E & \xrightarrow{f} & E \end{array}$$

Then $P(f)N(f) = N(\bar{f}) \cdot N(f_b)$.

PROOF. Recall the discussion at the beginning of this section. Lemma 4.8 tells us that $k_{ij} = P(f)$. Hence $N(\bar{f}) \cdot N(f_b) = \sum_{i=1}^{N(\bar{f})} \sum_{j=1}^{c_i} k_{ij} = P(f) \sum_{i=1}^{N(\bar{f})} c_i$ $= P(f) \cdot N(f)$. □

4.11 COROLLARY. If the fibration $q : E \to B$ is orientable and $\pi_1(E)$ and $\pi_1(F_b)$ are abelian, then Theorem 4.10 applies to every fiber map $f : E \to E$. □

A product formula involving $N_K(f_b)$

It is easier to relate $N(f)$ to $N(\bar{f}) \cdot N_K(f_b)$, since k_{ijK} is easier to compute than k_{ij}. A product theorem of this type is

4.12 THEOREM. Let $q : E \to B$ be orientable and suppose $\pi_1(E)$ is abelian. Then, for any fiber map $f : E \to E$, we have

$$P_K(f)N(f) = N(\bar{f}) \cdot N_K(f_b) ,$$

where $P_K(f)$ is the order of the quotient group

$$\frac{\mathrm{Fix}(H_1(B) \xrightarrow{\bar{f}_*} H_1(B))}{q_* \mathrm{Fix}(H_1(E) \xrightarrow{f_*} H_1(E))} .$$

PROOF. By the homotopy sequence of the fibration, we know $\pi_1(B)$ is abelian as well. Hence $\pi_1(E) = H_1(E)$, $\pi_1(B) = H_1(B)$, and by Theorem 1.6(iv), we know $k_{ijk} = P_K(f)$ for all i,j. Since $N_K(f_b) = \sum_{j=1}^{c_i} k_{ijK} = c_i \cdot P_K(f)$, when $b \in \bar{\mathbb{F}}_i$, it follows that c_i is independent of i. Now

$$N(\bar{f}) \cdot N_K(f_b) = N(\bar{f}) \cdot c_i \cdot P_K(f) = P_K(f) \sum_{i=1}^{N(\bar{f})} c_i = P_K(f)N(f) . \qquad □$$

HISTORICAL NOTES

Notes on the Introduction

The theorem on the torus was in [Nielsen (1921)] and [Brouwer (1921)], even earlier than the Lefschetz fixed point theorem (1923). For a modern treatment see [Brooks et al. (1975)].

Nielsen developed his theory of fixed point classes in [Nielsen (1927)] for surface homeomorphisms, using non-Euclidean geometry as a tool. Through the hands of Reidemeister and Wecken, it became a beautiful theory applicable to self-maps of polyhedra ([Wecken (1941)]). But computational results were few, see [Hopf (1927)] and [Franz (1943)]. It is interesting to note that Alexandroff and Hopf had planned to put Nielsen theory in their third volume (cf. [Alexandroff-Hopf (1935)] pp. 533-534 and 547-548), and that Leray, motivated by potential applications to analysis, had sketched a plan in his lecture at the 1950 International Congress of Mathematicians for extending the Nielsen theory to more general spaces. For the development of the subject in the 1960s see [Fadell (1970)] and [Brown (1974)].

Notes on Chapter I

Nielsen introduced the fixed point classes and the number bearing his name in his study of surface homeomorphisms [Nielsen (1927)]. It was not immediately clear how to generalize, because his method relies heavily on the fact that the universal covering of a surface of genus > 1 is the hyperbolic plane. Reidemeister gave a combinatorial treatment and considered the number bearing his name in [Reidemeister(1936)]. Then, pushing the covering space into the background and making very effective use of the alternative definitions 1.10 and 2.9, Wecken succeeded in generalizing the theory to continuous maps of polyhedra in his definitive work [Wecken I-III (1941)]. The extension to ANR's is due to [Browder (1948)]. The use of fat homotopies (Theorem 2.7) was suggested by [Scholz (1974)].

For the history of the theory of the fixed point index, the reader may consult [Brown (1971)].

The homotopy invariance theorem was proved in [Wecken (1941)]. The notion of the homotopy type of self-maps, the homotopy type invariance of the Nielsen number, and the characterization of the Nielsen number as the least number of fixed points in the homotopy type, was in [Jiang (1964)] and [Kiang-Jiang (1963)]. The commutativity theorem 5.2 appeared in [Fadell (1976)].

As for the equality between the Nielsen number and the least number of fixed points in a homotopy class, the first result was in [Wecken III]. He obtained Theorem 6.2, and a weaker form of Theorem 6.3 requiring X to be a manifold of dimension ≥ 3. [Shi (1966)] made considerable improvements. The present Theorem 6.3, along with Theorem 6.5, is in [Jiang (1980)]. It should be noted that [Shi (1975)] solved the computation of the least number of fixed points in the identity class, for an arbitrary compact polyhedron. The same problem for arbitrary homotopy classes is still open even for compact surfaces. But for homeomorphisms of closed surfaces, the Nielsen number is the least number of fixed points in the isotopy class [Jiang (1981)].

There are other views of the fixed point classes. See [Jiang (Sherbrooke 1980)], from differential topology, and [Fadell-Husseini (1981)], from obstruction theory.

There is a very promising algebraic approach to the theory of fixed point classes, started by [Reidemeister (1936)] and [Wecken II (1941)], which we do not touch in these lecture notes. For modern treatments see [Stallings], [Husseini] and [Grothendieck].

Notes on Chapter II

The exposition of §§1-5 follows the framework of [Jiang (1964)], in which the group $J(f)$ was introduced and Theorems 4.4 and 5.6 were proved. Theorems 2.3 and 2.5 appeared in [Kiang (1979)], and Theorem 4.9 in [Jiang (1979)]. Theorem 3.13 is due to [Gottlieb (1965)] and [Barnier (1967)]. Corollary 4.13 is also due to Gottlieb. It has an algebraic proof [Stallings (1965)]. Corollary 5.10 is taken from [Fadell-Husseini (1981)].

Converses of the Lefschetz fixed point theorem have been given in [Brown (1966, 1971)]. The present Theorem 6.1 is taken from [Jiang (1980)].

Notes on Chapter III

The notion of mod K fixed point classes was introduced by [Hirsch (1940)] and [McCord (1976)], primarily for the purpose of estimating the Nielsen number from below. They concentrated on the case of finitely-sheeted regular coverings, i.e. the case where $[\pi:K]$ is finite. A brief account can also be found in [You].

Section 3 is based on a paper by Jiang to appear in Pacific J. Math. I thank the journal for permission to use this material. The Nielsen type theory of periodic points started with [Halpern (1979)]. His results 4.14-15 were announced in [Halpern (1980)].

Notes on Chapter IV

The Nielsen theory for fiber maps originated with Brown (1967). The central problem was the product formula of Nielsen numbers. Through the years the subject developed in two directions. Pursuing the general theory, Fadell introduced the splittings bearing his name in [Fadell (1976)]. He also succeeded in generalizing the theory to compact ANRs. On the other hand, Pak introduced a generalized product formula in [Pak (1974)] with the correction factor bearing his name, under strong conditions on the fibration. The recent paper [You] gives a unified approach to both kinds of results, and simplified considerably the procedure of reducing the ANR case to the polyhedral case. Our exposition is based on this paper, though from a different viewpoint and with improvements (such as Theorem 4.2), and restricted to polyhedra for the sake of simplicity.

More specifically, the results 2.2, 4.4 and 4.5 are due to [Fadell (1976)], 2.8 is due to [Gottlieb (1968)], 3.2 is due to [Brown (1967)]. [Pak (preprint)] contains the conclusion of 4.10 under the stronger assumption that $\pi_1(E)$ is abelian and the fiber F satisfies the condition $J(F) = \pi_1(F)$. Theorem 4.1 and the present form of 4.10, as well as 1.6, 2.6, 3.4 and 4.12, are due to [You].

BIBLIOGRAPHY

Alexandroff, P., and Hopf, H. Topologie, Springer, Berlin, 1935.

Barnier, W. The Jiang subgroup for a map, Doctoral Dissertation, University of California, Los Angeles, 1967.

Bredon, G. E. Introduction to Compact Transformation Groups, Academic Press, New York, 1972.

Brooks, R., and Brown, R. A lower bound for the Δ-Nielsen number, Trans. AMS 143 (1969), 555-564.

Brooks, R., Brown, R., Pak, J., and Taylor, D. The Nielsen number of maps of tori, Proc. AMS 52 (1975), 398-400.

Brouwer, L. Über die Minimalzahl der Fixpunkte bei den Klassen von eindeutigen stetigen Transformationen der Ringflächen, Math. Ann. 82 (1921), 94-96.

Browder, F. E. The topological fixed point index and its application in functional analysis, Doctoral Dissertation, Princeton University, 1948.

__________. On continuity of fixed points under deformation of continuous mappings, Summa Brasil. Math. 4 (1960), 183-190.

__________. On the fixed point index for continuous mappings of locally connected spaces, Summa Brasil. Math. 4 (1960), 253-293.

Brown, R. F. On a homotopy converse to the Lefschetz fixed point theorem, Pacific J. Math. 17 (1966), 407-411.

__________. The Nielsen number of a fiber map, Annals of Math. 85 (1967), 483-493; Corrections, Annals of Math. 95 (1972), 365-367.

__________. Fixed points and fiber maps, Pacific J. Math. 21 (1967), 465-472.

__________. On the Nielsen fixed point theorem for compact maps, Duke Math. J. 36 (1969), 699-708.

__________. The Lefschetz Fixed Point Theorem, Scott-Foresman, Chicago, 1971.

__________. On some old problems of fixed point theory, Rocky Mountain J. Math. 4 (1974), 3-14.

Dold, A. Lectures on Algebraic Topology, Springer, 1972.

__________. The fixed point index of fiber-preserving maps, Invent. Math. 25 (1974), 281-297.

Eilenberg, S., and Steenrod, N. E. Foundations of Algebraic Topology, Princeton, 1952.

Fadell, E. Recent results in the fixed point theory of continuous maps, Bull. AMS 76 (1970), 10-29.

__________. Nielsen numbers as a homotopy type invariant, Pacific J. Math. 63 (1976), 381-388.

Fadell, E. Natural fiber splittings and Nielsen numbers, Houston J. Math. 2 (1976), 71-84.

Fadell, E., and Husseini, S. Fixed point theory for non-simply connected manifolds, Topology 20 (1981), 53-92.

Franz, W. Abbildungsklassen und Fixpunktklassen dreidimensionaler Linsenräume, Crelle J. 185 (1943), 65-77.

Geoghegan, R. The homomorphism on fundamental group induced by a homotopy idempotent having essential fixed points, Pacific J. Math. (to appear).

Gottlieb, D. H. A certain subgroup of the fundamental group, Amer. J. Math. 87 (1965), 840-856.

__________. On fiber spaces and the evaluation map, Annals of Math. 87 (1968), 42-55.

Grothendieck, A. Formules de Nielsen-Wecken et de Lefschetz en géometriè algébrique, (rédigé par I. Bucur), Exposé XII in Cohomologie ℓ-adique et Fonctions L, Lecture Notes in Math. vol. 589, Springer, 1977.

Halpern, B. Fixed points for iterates, Pacific J. Math. 25 (1968), 255-275.

__________. Periodic points on tori, Pacific J. Math. 83 (1979), 117-133.

__________. Periodic points on the Klein bottle, preprint.

__________. Nielsen type numbers for periodic points, preprint.

__________. The minimum number of periodic points, Abstract #775-G8, Abstracts AMS 1 (1980), 269.

Hirsch, G. Détermination d'un nombre minimum de points fixes pour certaines représentations, Bull. Sci. Math. 64 (1940), 45-55.

Hopf, H. Über Mindestzahlen von Fixpunkten, Math. Z. 26 (1927), 762-774.

Husseini, S. Y. Generalized Lefschetz numbers, preprint.

Jiang, B. J. Estimation of the Nielsen numbers, Acta Math. Sinica 14 (1964), 304-312 (= Chinese Math.-Acta 5 (1964), 330-339).

__________. Estimation of Nielsen numbers (II), Acta Sci. Natur. Univ. Pekin. 1979, 48-57.

__________. On the least number of fixed points, Amer. J. Math. 102 (1980), 749-763.

__________. Fixed point classes from a differential viewpoint, in Fixed Point Theory (Sherbrooke, 1980), Lecture Notes in Math. vol. 886, Springer, 1981.

__________. Fixed points of surfaces homeomorphisms, Bull. AMS 5 (1981), 176-178.

__________. On the computation of the Nielsen number, Pacific J. Math. (to appear).

Kiang, T. H. The Theory of Fixed Point Classes, Scientific Press, Peking, 1979 (English translation in preparation).

Kiang, T. H., and Jiang, B. J. The Nielsen numbers of self-mappings of the same homotopy type, Sci. Sinica 12 (1963), 1071-1072.

Leray, J. La théorie des points fixes et ses applications en analyse, in Proc. Intern. Congr. Math., Cambridge 1950, vol. 2, 202-208.

Massey, W. S. Algebraic Topology: An Introduction, Harcourt-Brace-World, New York, 1967.

McCord, D. An estimate of the Nielsen number and an example concerning the Lefschetz fixed point theorem, Pacific J. Math. 66 (1976), 195-203.

Nakaoka, M. Periodic points on nilmanifolds, preprint.

Nielsen, J. Über die Minimalzahl der Fixpunkte bei Abbildungstypen der Ringflächen, Math. Ann. 82 (1921), 83-93.

________. Untersuchungen zur Topologie der geschlossenen zweiseitigen Flächen, I, Acta Math. 50 (1927), 189-358; II, 53 (1929), 1-76; III, 58 (1932), 87-167.

________. Surface transformation classes of algebraically finite type, Math.-fys. Medd. Danske Vid. Selskab 21 no. 2 (1944), 1-89.

Pak, J. On the fixed point indices and Nielsen numbers of fiber maps on Jiang spaces, Trans. AMS 12 (1975), 403-415.

______. On the Reidemeister numbers and Nielsen numbers of fiber-preserving maps, preprint.

Reidemeister, K. Automorphismen von Homotopiekettenringen, Math. Ann. 112 (1936), 586-593.

Scholz, U. K. The Nielsen fixed point theory for noncompact spaces, Rocky Mountain J. Math. 4 (1974), 81-87.

Shi, G. H. On the least number of fixed points and Nielsen numbers, Acta Math. Sinica 16 (1966), 223-232 (= Chinese Math.-Acta 8 (1966), 234-243).

________. Least number of fixed points of the identity class, Acta Math. Sinica 18 (1975), 192-202.

Spanier, E. H. Algebraic Topology, McGraw-Hill, New York, 1966.

Stallings, J. Centerless groups -- an algebraic formulation of Gottlieb's theorem, Topology 4 (1965), 129-134.

Van der Walt, T. Fixed and Almost Fixed Points, Math. Centrum, Amsterdam, 1963.

Wecken, F. Fixpunktklassen, I, Math. Ann. 117 (1941), 659-671; II, 118 (1942), 216-234; III, 118 (1942), 544-577.

Whyburn, G. T. Topological Analysis, Princeton, 1964.

You, C. Y. Fixed point classes of a fiber map, Pacific J. Math. (to appear).

INDEX

LIST OF SYMBOLS

ABCDEFGHIJ–AMS–89876543